# EINSTEIN'S ITALIAN MATHEMATICIANS

## RICCI, LEVI-CIVITA, AND THE BIRTH OF GENERAL RELATIVITY

# EINSTEIN'S
# ITALIAN MATHEMATICIANS

## RICCI, LEVI-CIVITA, AND THE BIRTH OF GENERAL RELATIVITY

Judith R. Goodstein

AMS
AMERICAN MATHEMATICAL SOCIETY
Providence, Rhode Island

2010 *Mathematics Subject Classification.* Primary 01-02, 01A55, 01A60, 01A70, 83-03.

For additional information and updates on this book, visit
**www.ams.org/bookpages/mbk-113**

**Library of Congress Cataloging-in-Publication Data**

Names: Goodstein, Judith R., author.
Title: Einstein's Italian mathematicians : Ricci, Levi-Civita, and the birth of general relativity / Judith R. Goodstein.
Description: Providence, Rhode Island : American Mathematical Society, [2018] | Includes bibliographical references and index.
Identifiers: LCCN 2018000022 | ISBN 9781470428464 (alk. paper)
Subjects: LCSH: Levi-Civita, Tullio, 1873–1941. | Ricci Curbastro, Gregorio, 1853–1925. | Mathematicians–Italy–Biography. | Von Kármán, Theodore, 1881–1963. | Einstein, Albert, 1879–1955. | Calculus of tensors. | General relativity (Physics)–Mathematics. | AMS: History and biography – Research exposition (monographs, survey articles). msc | History and biography – History of mathematics and mathematicians – 19th century. msc | History and biography – History of mathematics and mathematicians – 20th century. msc | History and biography – History of mathematics and mathematicians – Biographies, obituaries, personalia, bibliographies. msc | Relativity and gravitational theory – Historical (must also be assigned at least one classification number from Section 01). msc
Classification: LCC QA28 .G66 2018 | DDC 510.92/245–dc23
LC record available at https://lccn.loc.gov/2018000022



∞ The paper used in this book is acid-free and falls within the guidelines
established to ensure permanence and durability.
Visit the AMS home page at `https://www.ams.org/`

10 9 8 7 6 5 4 3 2 1     23 22 21 20 19 18

For David, Marcia, and Mark

# Contents

# Preface

This tale of the Italian mathematicians who provided Albert Einstein with the crucial mathematical architecture for general relativity might never have come about had it not been for a piece of mail that surfaced more than forty years ago in the correspondence of a very different mathematician, the Hungarian-born polymath Theodore von Kármán. He spent many years on the faculty of the California Institute of Technology, and early in my career as Caltech's first archivist, while rummaging through a collection of his letters, I came across one that he had received in 1922 from Tullio Levi-Civita, a mathematics professor in Rome. The name was unfamiliar to me, but even before reading the letter, I was struck by the handwriting. Unlike many scientists whose scrawls I had alternately puzzled over and despaired of, Levi-Civita had mastered the art of penmanship. His script was distinguished by bold, slightly exaggerated letters, with every I dotted and T crossed. His prose marched in straight, confident lines across the page, all correctly punctuated. What he had to say was at least as interesting as his handwriting: In the bitter aftermath of World War I, the victorious Allied nations, including Italy, had actively discouraged their scientists from attending meetings with their counterparts in the defeated nations of Germany and Austria-Hungary. Von Kármán, then still teaching in Germany, had evidently written to Levi-Civita to ask for his help in lifting the embargo. In his response, Levi-Civita had offered his enthusiastic support, now that "the brutal parenthesis of the war" had ended, for an international conference von Kármán proposed holding in Innsbruck, Austria. "Almost" all of his Italian colleagues would attend, he assured him, "without deplorable lingering of a bellicose mentality" still sometimes displayed by their peers.

Intrigued by the tone of the letter, I decided to investigate further. I quickly discovered that Levi-Civita had been perhaps the most brilliant and versatile member of a circle of influential Italian mathematicians—a number, like him, from Jewish backgrounds—that had flourished in the decades after the unification of Italy in 1870, only to be undone by the collapse of the nation's democratic institutions following the rise of Mussolini and fascism in the years leading up to World War II. He himself had died in 1941, stripped of his academic posts and prerogatives under the state's anti-Semitic laws and ostracized by colleagues who only a few years earlier had hailed him as their field's shining light.

My language skills did not extend to Hungarian and my German wasn't much better, but I could read and, after a fashion, speak Italian. Moreover, I would be accompanying my physicist husband, David, on his research trip to Rome in the near future. Putting aside for the moment any further inquiries into von Kármán, I found myself not long afterward standing in front of Via Sardegna 50, Levi-Civita's last known address in Rome. There I encountered a kindly janitor who told me that Levi-Civita's widow, Libera, had recently moved to another part of the city. While I stood in the empty apartment, he called her and handed me the phone. In my halting Italian, I introduced myself and explained my interest, and she invited me to visit her, which I did about a week later.

When I met Libera for the first time that summer, I knew painfully little about her and not that much more about her distinguished husband. Considering that she knew absolutely nothing about me, her subsequent generosity was astonishing. We talked for a while and then she directed me to an armoire located in an alcove adjacent to her living room, and said, "His letters are here." I already knew Levi-Civita to be a tireless correspondent, whose letters, written between 1896 and 1941, had been promised to Italy's Lincei Academy of Sciences, but I had not expected to find them tucked away in a corner of her home, numbering in the thousands.

A few weeks later we met again, and this time she introduced me to her son in law, Pier Vittorio Ceccherini, a professor of mathematics in Rome who had been a student of Beniamino Segre, now the president of the Lincei. The three of us came to an understanding that I could examine some of the letters while I was in Rome and that the Lincei would mail the complete collection to me in Pasadena, California, the following summer. They arrived still tucked in their envelopes, bundled together by year, each packet neatly tied with string. Over the next several months, I pored over their contents, consulted other sources, and began assembling a picture of Levi-Civita's life, his research, and the intellectual milieu in which he lived and worked. I discovered that he and his mathematical peers had assumed the primary role in disseminating relativity theory throughout Italy's scientific community, and that Levi-Civita himself had carried on a lively correspondence with Einstein in the years immediately before and after World War I. I presented the early fruits of this research in an invited talk at the Lincei Academy in 1973 and over the next decade published a handful of additional articles about Italy's nineteenth-century renaissance in mathematics.

Years later, having completed a book about the history of Caltech, I returned to the themes of those earlier pieces and expanded them into a book *The Volterra Chronicles,* which was published in 2007. It was sometime afterward that Sergei Gelfand, my shrewd and knowledgeable publisher at the American Mathematical Society, said to me, "What about Ricci?" Well, what about Ricci? He was certainly not one of the flashier figures of that era. Thinking back to Levi-Civita's letters, I vaguely recalled that Gregorio Ricci had been his professor and later his colleague at the University of Padua.

Their subsequent relationship had been close enough, I soon discovered, for Levi-Civita to have delivered an eloquent and emotional tribute to him at the Lincei in 1926, a year after his death. I located a copy of that talk (which is translated into English for the first time here in its entirety), read its account of the indispensable role Ricci's absolute differential calculus had played in the formulation of general relativity, and embarked on a journey that has brought me here.

The premise of this book is simply this: that the part Ricci and Levi-Civita played in contributing to Einstein's theory is a story that deserves to be told and told to a wider audience than it has reached to date. Recognizing Gregorio Ricci for his role in this saga—a mathematician who spent his whole adult life engaging in what the mathematical physicist Freeman Dyson has called "unfashionable science," but who also had the extraordinary good fortune to encounter Tullio Levi-Civita as an effervescent student at Padua and form an enduring friendship with him—has been the driving force behind this narrative. Einstein, of course, features in this drama too—not the central figure but still a major player whose sudden, somewhat disheveled, appearance in the third act throws into sharp relief the role of its principals in setting the stage for one of the great, transformative discoveries of twentieth-century science.

**Acknowledgments.** This book owes much to Pier Vittorio Ceccherini and Susanna Silberstein Ceccherini, who first opened their home to me more than forty years ago and permitted me unfettered access to Levi-Civita's correspondence and private documents and other manuscript material in their possession. Tullio Ceccherini-Silberstein, their son, also a mathematician, shared family photographs with me and read the chapters as did my colleagues Tilman Sauer and Michele Vallisneri, who rescued me from time to time with their mini-tutorials on relativity and tensors. Donald Babbitt, who has been my mentor in all things mathematical, has been a loyal friend and staunch supporter for the past decade.

The publisher of the American Mathematical Society, Sergei Gelfand, made me an offer I couldn't refuse in suggesting that Ricci deserved to be brought out of the shadows. He challenged me to answer the question: "Who was he?" In 2004, Fabio Toscano published an Italian-language book about Ricci and Levi-Civita, which I found helpful.

While the considerable archival research undertaken in the preparation of this book is amply reflected in the footnotes, the kindness and generosity of the many archivists and librarians who facilitated my work deserve a special thank you In particular, I would like to thank Orith Or Burla, Chaya Becker, and Roni Grosz, all of the Albert Einstein Archives at The Hebrew University of Jerusalem, for their help and permission to quote from The Collected Papers of Albert Einstein. In the early stages of writing this book, I was a visiting associate at the Einstein Papers Project at Caltech, courtesy of Diana Kormos Buchwald, the project's general editor, who provided immense help with her knowledge of the papers and patiently answered many questions.

In Rome, Francesca Rosa provided many hours of much-needed research help in the Archivio Centrale dello Stato; the Archivio di Stato; the Archivio Storico and the Physics Department of the University 'La Sapienza'; the Archivio Storico del Senato della Repubblica; and the Accademia Nazionale dei Lincei. I am grateful for the help she received from Mariapina Di Simone, Carla Onesti, Marco Guardo, Susanna Panetta, and Andrea Dibitonto. In Pisa, my thanks to Diana Toccafondi, Giovanna Tanti, Daniela Staccioli, Damiano Cecchi, Sandro Di Majo, Maddalena Taglioli, Anita D'Orazio, Mauro Bellandi, and Alessandro Masi at the Archivio di Stato, the Scuola Normale, and the Università di Pisa, who answered countless questions, supplied copies of documents, and in some cases photographs. I thank all of them for their help and permission to quote from materials in their repositories.

In Lugo di Romagna, Luisa Bedeschi and Ivana Pagani provided friendship, enthusiasm, unlimited scans, and many helpful suggestions as Carlotta Scaramuzzi, who translated many of the documents, and I delved into the Ricci papers. Enrico Sangiorgi, the great-grandson of Gregorio Ricci, accompanied by his aunt, Anna De Rossi, organized a visit to Ricci's country house in nearby Sant'Agata sul Santerno, including a reception in the town's library offered by Armanda Capucci, the librarian and Enea Emiliani. The high point of the day for me was a tour of the house itself, which gave me a real sense of Ricci's way of life there.

In Padua, I had a lively correspondence with Mariarosa Davi, a teacher at the prestigious Liceo Ginnasio Tito Livio, who went beyond the call of duty to track down and send me Tullio Levi-Civita's pre-university academic records. Across the Alps, Bärbel Mund in Gottingen provided help in locating Ricci correspondence in the Felix Klein Nachlass, and Heike Hartmann offered assistance in selecting pictures in the ETH's photographic archive.

Closer to home, my appreciation goes to the document delivery staff at Caltech's Sherman Fairchild Library, who never failed to find the pamphlets, articles, and books I needed. A big thanks also to Loma Karklins and Elisa Piccio in the Caltech Archives who gave their time and help to my book-writing project.

Translations from one language to another can be unusually challenging at times. I enlisted the help of my husband, David, and our daughter, Marcia Goodstein, in translating several difficult documents from the Italian; Donald Babbitt did the same for French material. James T. Smith provided a valuable preliminary translation of Levi-Civita's memoir about Ricci's life and work.

Various friends have read the chapters and provided thoughtful and constructive suggestions.

I owe a special debt of gratitude to Heidi Aspaturian, my long-time editor and colleague at Caltech, who through many detailed and discerning readings of the chapters helped turn the manuscript into a compelling story. As the saying goes, behind every writer is a better editor and in my case, certainly, that statement has never been truer or more appreciated.

Judith Goodstein
July 2017

# Permissions & Acknowledgments

The American Mathematical Society gratefully acknowledges the kindness of the following individuals and institutions in granting permission to reprint material in this volume:

**Accademia Nazionale dei Lincei**

English translation of *Commemorazione del socio nazionale prof. Gregorio Ricci-Curbastro*, Memorie della Classe di scienze fisiche, matematiche e naturali, s. VI, vol. **1** (1926), pp. 555–564.

**Photo 7**. Eugenio Beltrami (1835–1900). Credit: Archivio dell'Accademia Nazionale dei Lincei, Fondo Vito Volterra.

**Photo 9**. Ernesto Padova (1845–1896). Credit: Archivio dell'Accademia Nazionale dei Lincei, Fondo Vito Volterra.

**Photo 12**. Gregorio Ricci Curbastro (1852–1925). Credit: Archivio dell'Accademia Nazionale dei Lincei, Fondo Vito Volterra.

**Photo 17**. Felix Klein (1849–1925). Credit: Archivio dell'Accademia Nazionale dei Lincei, Fondo Vito Volterra.

**Photo 19**. Hermann Minkowski (1864–1909). Credit: Archivio dell'Accademia Nazionale dei Lincei, Fondo Vito Volterra.

**Archivio delle Ancelle Sacro Cuore di Lugo**

**Photo 1**. Ricci Curbastro family tree.

**Archivio Fotografico, Biblioteca Comunale Ricci Garotti, Sant'Agata sul Santerno, RA, Italia**

**Photo 30**. Gregorio Ricci Curbastro and family, Sant' Agata sul Santerno (ca. 1921).

**Biblioteca Comunale Fabrizio Trisi, Lugo**

**Photo 2**. Gregorio Ricci Curbastro (ca. 1869). Credit: Ida Ricci Curbastro/Biblioteca Comunale Fabrizio Trisi.

**Photo 3**. Postcard scene, Lugo (ca. 1900).

**Photo 10**. Bianca Bianchi Ricci Curbastro. Credit: Maria Giovanna Ricci Curbastro/Biblioteca Comunale Fabrizio Trisi, Lugo.

**California Institute of Technology**

**Photo 22**. Title page of the "Outline [or preliminary version] of a Generalized Theory of Relativity and of a Theory of Gravitation." Credit: Paul Epstein Papers, Courtesy of the Archives, California Institute of Technology.

**Photo 24**. Max Abraham (1875–1922). Credit: Courtesy of the Archives, California Institute of Technology.

**Photo 31**. Postcard from Levi-Civita to Theodore von Kármán, 1921. Credit: Courtesy of the Archives, California Institute of Technology.

**Ceccherini-Silberstein Family**

**Photo 14**. Signed portrait of Garibaldi.

**Photo 23**. Tullio Levi-Civita (1873–1941) (ca. 1912).

**Photo 28**. Passport for Libera Levi-Civita, issued in Padua, Sept. 30, 1918.

**Photo 29**. Gregorio Ricci Curbastro (1852–1925).

**Photo 32**. Tullio Levi-Civita, La Plata, Argentina, 1923.

**Enrico Persico Papers, Dept. of Physics, Sapienza University, Rome**

**Photo 33**. A humorous depiction of Levi-Civita revealing the mysteries of the absolute differential calculus (ca. 1923).

**ETH-Bibliothek Zürich/Bildarchiv**

**Photo 20**. Marcel Grossmann (1878–1936). Credit: ETH-Bibliothek, Bildarchiv/Fotograf: Unbekannt/Portr_01239/Public Domain.

**Photo 21**. Albert Einstein (1879–1955). Credit: ETH-Bibliothek, Bildarchiv/Fotograf: Unbekannt/Portr_03101/Public Domain.

**Fondo 'Ricci Curbastro,' Il Liceo Statale 'Ricci Curbastro' di Lugo (RA), Italy, III Raccoglitore**

**Photo 8**. Pages from the manuscript for Ricci's inaugural lecture in Padua, 1881. Credit: Fondo 'Ricci Curbastro,' Il Liceo Statale 'Ricci Curbastro' di Lugo (RA) Italy, III Raccoglitore, #48.

**From a Private Collection**

**Photo 4**. Palazzo della Sapienza.

**Photo 27**. Caffè Pedrocchi, Padua (ca. 1915).

**Photo 34**. Tullio Levi-Civita and his wife Libera Trevisani.

**The Hebrew University of Jerusalem.**

**Photo 25**. Pages from a Levi-Civita letter to Einstein, March 28, 1915. © The Hebrew University of Jerusalem, Israel.

**Photo 26**. Two postcards from Einstein to Levi-Civita, April 14 and April 21, 1915. © The Hebrew University of Jerusalem.

**Ida Ricci Curbastro**

**Photo 2**. Gregorio Ricci Curbastro (ca. 1869). Credit: Ida Ricci Curbastro/Biblioteca Comunale Fabrizio Trisi.

**Institut Mittag-Leffler**

**Photo 6**. Ulisse Dini (1845-1918).

**Maria Giovanna Ricci Curbastro**

**Photo 10**. Bianca Bianchi Ricci Curbastro. Credit: Maria Giovanna Ricci Curbastro/Biblioteca Comunale Fabrizio Trisi, Lugo.

**Photo 11**. Bianca Bianchi Ricci Curbastro.

**The Royal Society**

**Appendix C**. "Tullio Levi-Civita" by W. V. D. Hodge is reprinted with permission of the publisher, The Royal Society, from *Obituary Notices of Fellows of the Royal Society of London*, **4** (1942), 151–165.

**Scuola Normale Superiore Pisa**

**Photo 5**. Enrico Betti (1823–1892).

**Senato della Repubblica**

**Photo 13**. Giacomo Levi-Civita (1846–1922). Credit: ASSR, Senato del Regno, Segreteria, fascicolo personale del senatore Giacomo Levi-Civita, n. 1274.

**Wikimedia Commons**

**Photo 15**. Giuseppe Veronese (1854–1917).

**Photo 16**. Luigi Bianchi (1856–1928).

**Yale University**

**Photo 18**. Ricci-Levi-Civita's paper published in *Mathematische Annalen* **54** (October, 1900). Credit: Cushing/Whitney Medical Library, Yale University. Public Domain.

CHAPTER 1

# The Riccis of Lugo

Gregorio Ricci Curbastro was born on January 12, 1853, in Lugo di Romagna, a small papal town of about 22,000 inhabitants, on the railway line between Bologna and Ravenna, in the northeast corner of the Italian peninsula. The custom of adding the name of a wife or mother to the Ricci surname dates back many centuries, and was likely adopted to differentiate the various branches of Ricci's forefathers in that region. Ricci himself used only the first part of his name in the landmark 1900 tensor analysis paper he coauthored with Tullio Levi-Civita—and he has been known by that single name ever since.

He was the second of four children. (Two younger sisters went to convent schools and later became nuns. His older brother, Domenico, became a civil engineer.) Their mother, Livia Vecchi, the daughter of a prominent engineer in the province of Bologna, was a devout Roman Catholic whom Ricci adored. A stickler for observing the catechism of the Church and a zealous guardian of "faith and Christian morals," she practiced Christian charity and humility every day of her life, so Ricci described her.[1] He was, perhaps, even more emotionally attached to his sister, Giustina (later Sister Vincenza), whom he saw as the image of his mother.

His father, Antonio Ricci Curbastro, was an engineer—"a man of great piety and a Catholic of the old school,"[2] a contemporary later recalled—and a substantial landowner who fed the poor and gave generously to the Church. The Ricci family had a long and politically active history in the region, an aristocratic pedigree, and numerous business interests, which Antonio managed. He also served for a time as the pope's *gonfaloniere,* Lugo's highest-ranking church official. In 1857, on the one occasion that Pius IX visited Lugo, he spent the night at the Ricci family palazzo.

Two years after the papal visit, the country's long political struggle for independence erupted in Ricci's backyard. Before the *Risorgimento,* the peninsula's campaign for independence and national unification, Italy consisted of a hodgepodge of states, from duchies big and small, to the kingdoms of Sardinia and the Two Sicilies, to the states of the Church—all of them owned, traded, and fought over by a succession of foreign rulers and popes. Then, between 1859 and 1861, Camillo Cavour, the prime minister of the northwest region of Piedmont, systematically annexed one region after another, using popular uprisings, rigged plebiscites, and military intervention to transform "a geographical expression" (as the Austrian statesman Metternich had once characterized Italy) into a single nation under one ruler. With the proclamation of the Kingdom of Italy in 1861, Victor Emmanuel (Vittorio Emanuele) II of the House of Savoy spoke in the name of twenty-two million people, four percent of whom had the right to vote. Only Venice, Rome, and Trieste remained beyond his grasp, until they too joined the new nation in 1866, 1870, and 1919 respectively.

As children, Ricci and his siblings joined their mother as she strolled up and down the streets of Lugo. In later life he remembered these outings "as always far less than pleasant because of all the long and repeated pauses" as his mother lent a sympathetic ear to the poor women of the town. Listening patiently to their litany of personal problems, she always offered "some help or comfort," he recalled, adding, "In spite of that, she preserved such a natural dignity, that more than lowering herself, it looked like she was raising those poor women to her level."[3] Ricci's character, like his mother's was shaped by deeply held religious beliefs and a fierce perseverance, traits he carried into adulthood.

Starting at age nine, Ricci and his brother Domenico, who was two years older, learned their three R's at home. In 1864, after two years of Italian and Latin instruction under the guidance of a retired grammar teacher, Don Savino Poletti, a new teacher, Francesco Taglioni, tutored them in the humanities and rhetoric for the next five years. The future mathematician, Taglioni reported, brought to his studies a combination of qualities he had rarely found in a student: a "penetrating mind, a lively ingenuity, a tenacious memory, and what one searches for, and that matters, in a pupil, a desire to learn."[4] Adept in all subjects, Ricci shone in Latin and Greek and in translating long passages from Cicero, Virgil, and Homer into elegant Italian prose, sprinkled with poetic flourishes. From the world of classical literature, teacher and pupil moved into the realm of philosophy, where Ricci excelled at defending "philosophical truths and exposing false systems with fresh common sense considerations and arguments." Two years of private instruction in mathematics, physics, and astronomy, taught by Giuseppe Manzieri, a professor of physics and mathematics at the local high school, rounded out Ricci's secondary school education.

That Ricci would continue his studies at an Italian university was never in doubt. While Bologna and Padua were both venerable and well-established institutions of higher learning, family tradition argued for the Archiginnasio Romano, the papal university in Rome, as both Antonio and Livia's fathers had studied there. In fall 1869, armed with glowing letters of recommendation from his tutors, Gregorio and Domenico (also a solid student) boarded the train for Rome, the last of the papal cities, where the younger brother intended to pursue a degree in mathematics at the ancient University of Rome, which also offered instruction in medicine, law, and theology. Founded by Pope Boniface VIII at the beginning of the fourteenth century,[5] the papal university (popularly known as La Sapienza) was housed in an austere palazzo consisting of a massive two-story colonnade wrapped around an elegant internal courtyard with its own house of worship, the Church of Sant'Ivo, designed by the Baroque architect Francesco Borromini, at one end of the courtyard. The two young men rented lodgings within shouting distance of the papal residence of Pius IX.

Rome in 1869 was a bustling city of competing political alliances, monumental architecture, and sharp social contrasts. The French troops of Napoleon III, garrisoned in the city since 1864, patrolled the streets and guarded the pope. While Prime Minister Cavour regarded Rome as the future capital of Italy, King Victor Emmanuel, acting on his own, was inexplicably seeking an alliance with Napoleon that would have prevented his own troops from laying claim to Rome and the pope's temporal powers. Nor did the local populace seem inclined to rise up and send the pope's mercenaries packing. As the distinguished English historian Denis Mack Smith has observed, "Rome was still the 'parasite city' of clerics, hotelkeepers, and beggars." To compound the matter, he adds, "Half of its population existed on official hand-outs, and the governing classes of clergy exempted themselves from taxes and had a highly privileged position to defend."[6]

Soaring columns of marble and decaying arches from the time of the Caesars beckoned in one quarter of the city; in another, the Corso, a long street terminating at the grand Piazza del Popolo, the "People's Square," bustled with massive palaces, stores, and private homes. Innumerable churches, often built on the foundations of ancient Roman temples, could be found on every block. Numerous street festivals, fireworks from the bulwarks of the Castel Sant'Angelo—used as a papal fortress since medieval times—and even the public beheadings of common criminals brought out a crowd.

The saddest quarter of the Eternal City was reserved for Rome's Jewish population, some 3,800 souls who lived in the ghetto built on the banks of the Tiber River, which often overflowed. In *Pictures from Italy,* written while he was traveling through the peninsula in 1844, English author Charles Dickens painted a bleak picture of the daily lives of the ghetto's inhabitants. "The little town of miserable houses," he wrote, after surveying this part of the city one moonlit night, "walled, and shut in by barred gates, is the quarter where the Jews are locked up nightly, when the clock strikes eight—a miserable place, densely populated, and reeking with bad odors, but where the people are industrious and money-getting." Wandering in the lanes and alleys during the daylight hours, he added, "you see them all at work; upon the pavement, oftener than in their dark and frowzy shops: furbishing old clothes, and driving bargains."[7] From the time of Dickens's visit to Ricci's freshman year twenty-five years later, conditions in the ghetto remained unchanged.

On account of Ricci's age—he was a couple of months shy of eighteen, the Roman university's age requirement for matriculation—his academic year began with an undated petition to Cardinal Karl-August von Reisach, prefect of the sacred congregation of studies, requesting admission to enroll in the school's course in philosophical and mathematical studies. Attached to Ricci's request were various letters attesting to his character and academic ability, which he had dutifully brought with him from Lugo. Duly granted admission, Ricci registered for calculus, complementary algebra and analytic geometry, higher philosophy and physics, won accolades from his instructors at the end of each quarter, and in June 1870, left Rome, with his bachelor's diploma, in mathematics. This was a certificate that the school conferred on students who had successfully passed an examination in their field at the end of the first year's work. Third and fourth year students who passed their final exam earned a diploma (*la licenza*) and a doctor's degree (*la laurea*) respectively.

Ricci was back home in Lugo when war between the armies of Napoleon III and Bismarck broke out in July 1870, ending two months later in a crushing defeat for the French at the hands of the Prussians. Bismarck's victory decisively altered the balance of power in Europe and provided King Victor Emmanuel with the pretext he needed to occupy Rome. In the absence of the French troops, who had left Rome that August for the front lines, the king's troops broke through the ancient walls of the city at Porta Pia on September 20, bringing an end to papal rule in Rome, emancipating the city's Jews, and laying the groundwork for the city's annexation to the kingdom of Italy. A plebiscite held that October overwhelmingly approved of the invasion, thus completing the unification of Italy. Overnight, and with barely any armed resistance inside the walls, Rome had become the nation's capital.

In the fall of 1870, Gregorio and Domenico returned to Rome to continue their university studies. However, the collapse of the pope's temporal power in Rome had dramatically changed the city's academic landscape. The Archiginnasio Romano had relocated to a private palace, along with faculty and one hundred and twenty students who remained loyal to the pope. Meanwhile, the new occupant of the Palazzo della Sapienza, the Reale Università degli Studi di Roma—the imprimatur of the king emblazoned in its very title—had taken the first step in its quest to become the crown jewel in Italy's pantheon of state-supported educational institutions. Although the documentation is sketchy on this point, one of the Ricci brothers, not identified, fell sick before the fall term began, and both of them subsequently missed the first month of school. Afterwards, they attended classes, but as auditors, not registered students. No record of their attendance that year has come to light. Presumably, they continued to frequent the pontifical university, which remained in clerical hands until 1876 when the minister of public instruction, issued a decree suppressing the university.

Before 1870 ended, the brothers had once again returned to Lugo, intending to transfer to the University of Bologna the following fall. They were stunned to be refused admission on the grounds that Bologna, among other criteria, required its students to have high-school diplomas, which they lacked. When an appeal to the ministry of public instruction fell on deaf ears, the family approached Gaspare Finali, an influential politician who had played a prominent role in the referendum to end papal sovereignty in the province of Romagna in 1859. A consummate political insider who seemed to know everyone connected with the government, he immediately wrote to Cesare Correnti, the minister of public instruction, to raise the question of fairness in dealing with transfer students: "They [the Ricci brothers] ask that the years spent studying in Rome be taken into account: Is this not conceded to the students coming from foreign universities, provided that they take the special exams for those areas of science not covered by the program of the university they come from?" Finali suspected, probably correctly, that the problem of matriculation at Bologna had more to do with the poor quality of the education the brothers had received in Rome than anything else. "I don't think that the ministry wants to use higher standards for two young men coming from a national university, even if it has not yet been brought to national standards; they are genuinely talented and studious young people, and we should facilitate their passage, without doing damage to them.... I beg you to let me know how the ministry is prepared to resolve the request by the Ricci brothers; and if this current request should not succeed, I beg you in the name of the regard that you have for aspiring young men, to tell me in what way and by what means they should present it."[8]

As tactfully as possible, Finali had also pointed out to Correnti that their mutual colleague Luigi Luzzatti, secretary general to the minister of agriculture, industry, and commerce, offered students an alternative route to obtaining the equivalent of a high school diploma. When Ricci subsequently learned that a professional diploma from one of Italy's technical institutes would satisfy Bologna's entrance requirements, he wasted no time pursuing this course of action. In summer 1871, he took—and passed—the mandated examination for a physics-mathematics teacher and received a diploma from the Technical Institute of Bologna, signed by Luzzatti.

In November 1872, now sporting two diplomas, Ricci again requested permission to enroll at Bologna as a second-year student in pure mathematics. This time he was admitted, on the condition that he also attend the university's lectures in inorganic chemistry and design and take the year-end final exams in these courses, along with his regular courses. In a letter he wrote to a cousin shortly after beginning his classes, Ricci summed up the main reason for abandoning his studies in Rome in four words: "the events of 1870." Those events included the prospect of instruction in a new building without any labs, the dearth of functioning public offices, a huge influx of newcomers in need of housing, street demonstrations, and labor strikes, capped by "the devastating flood [whose waters rose more than seventeen meters above sea level in the low lying parts of the city] in the final days of 1870, which extensively disrupted city life."[9] At the same time, his months at home had given him a taste of what lay in store for his birth city—and it filled him with anticipation.

The proximity of Bologna to Lugo, he wrote, "means I can follow more attentively and with more zeal the political reforms that our countrymen who have created a united Italy are going to carry out, among them Renzo [a prominent local official], who has played such a major part both in restoring the finances of the town and making Lugo more beautiful."[10]

Ricci spent exactly one year at Bologna (Domenico remained there and subsequently graduated with a degree in engineering). In the first weeks of July 1873, he took the lengthy oral exams, administered by a trio of examiners. In inorganic chemistry, differential and integral calculus, and descriptive geometry, the examiners awarded him 30—the highest possible grade—with praise, while in physics he achieved a more-than-respectable mark of 27. Despite this auspicious beginning, he left Bologna that fall for Pisa. Attrition within the department may have been a factor. One of his examiners, Eugenio Beltrami, a prolific differential geometer and mathematical physicist, who had served as one of his examiners in descriptive geometry, decamped to Rome at the same time to take up new duties as professor of rational mechanics. Six years earlier, Luigi Cremona, who studied the birational transformations of plane and space ("the Cremona transformations"), had left Bologna for the Polytechnic Institute of Milan. Beltrami and Cremona took much of the luster and prestige of the department with them, casting a long shadow over its reputation until around 1880 when Bologna rebuilt its mathematics program with a new generation of teachers, researchers, and students. Ricci and Beltrami would cross paths again in 1887 when Ricci competed unsuccessfully in a national competition for the Royal Prize for Mathematics established by Umberto I, who succeeded Victor Emmanuel as king in 1878.

When Ricci had entered Bologna in 1872, he lacked academic respectability in the eyes of that university's gatekeepers. When he enrolled at the Scuola Normale Superiore in Pisa the following year as a third-year transfer student, he brought with him unimpeachable scholarly credentials and an air of reserve that he never lost. The mathematicians in Pisa who zealously guarded the reputation and intellectual integrity of both their courses and their institution embraced the brilliant and diffident young man and set him on the road that would lead to the invention of the absolute differential calculus.

CHAPTER 2

# The Making of a Mathematician

Unlike the University of Bologna, the Scuola Normale Superiore di Pisa offered Ricci what he needed most in 1873—a thriving school of mathematics connected to the University of Pisa. The school, which had flourished briefly (1810–1815) as a branch of the French École Normale Supérieure in Paris, following the annexation of Tuscany to Napoleon's empire, closed after Napoleon's defeat at Waterloo. It reopened in 1847, under the patronage of Leopold II of Lorena, in the Palazzo Carovana, Giorgio Vasari's Renaissance masterpiece, once upon a time the headquarters of the knights of Santo Stefano. Between the proclamation of the Kingdom of Italy in 1861 and the end of World War I the school burnished its reputation as an elite institution devoted to training teachers in letters and philosophy and in the physical and mathematical sciences. In their definitive history of the Scuola Normale, Tina Tomasi and Nella Sistoli Paoli single out Enrico Betti, Ulisse Dini, Eugenio Bertini, and Luigi Bianchi for their roles in creating distinctive Italian schools of mathematical physics, analysis, and algebraic and differential geometry.

Betti, the school's director between 1865 and 1900, was a true son of the *Risorgimento.* As a young man, he had fought in the battles of Curtatone and Montanara during the first Italian war of independence in 1848 (the Austrian army prevailed against the Tuscany university battalion led by Mossotti, Betti's thesis advisor at Pisa). Early in his career, while teaching high school in Pistoia and Florence, he concentrated on the resolution of algebraic equations, which led to an appointment in 1857 as professor of higher algebra at the University of Pisa. At Pisa Betti had compiled an impressive academic record as teacher, writer, and mathematician in the scientific world. Behind him lay his early work on the resolution of algebraic equations (extending and furnishing one of the first rigorous proofs for the work of Galois), the theory of elliptic functions (the first Italian treatise on the subject), and the translation into Italian of Riemann's doctoral thesis (completed in 1851 under the supervision of Gauss), "Foundations for a General Theory of Functions of a Complex Variable," for the *Annali di Matematica pura e applicata* (*Annals of Pure and Applied Mathematics*), a new journal geared to the needs of a new generation of Italian mathematicians, among them the young Ricci.

Of the ten students competing for one of the Scuola Normale's coveted spots in the physical and mathematical section in 1873, the examiners rejected five of them. Ricci took—and passed with a score of 35 out of a possible 40—the rigorous oral and written examination for admission as an external student to the school, meaning he did not live in the Palazzo Carovana, as did some of the other students. Luigi Bianchi, whose work in differential geometry would later contribute in a fundamental way to the construction and understanding of the general theory of relativity, was also admitted as a Normalista (the colloquial term for a Scuola student) that November. The examiners included Betti, who by then had formally relinquished his chair of analysis and geometry in favor of the chair of celestial mechanics at Pisa, and Dini, who had been Betti's student. Both had been instrumental in guiding the school to its preeminent position in mathematical research and teaching after Italy's unification. Ernesto Padova, another of Betti's top doctoral graduates, who later became Ricci's colleague and good friend at the University of Padua, served as the third examiner.

The academic record of Ricci's coursework, preserved in the archives of the school, indicates that he spent the first year juggling four required math and science classes at the university with four classes devoted to higher geometry and differential and integral calculus at the Scuola Normale. A German language class rounded out his course schedule. He took Paolo Tassinari's course in organic chemistry and Giuseppe Meneghini's in geology and mineralogy. His other professors at Pisa that year included Padova, who taught rational mechanics, Cesare Finzi, an accomplished teacher who lectured on differential and integral calculus, and Dini, who introduced him to mathematical analysis and higher geometry, a subject close to Dini's heart. Dini, in fact, filled in for Betti as director of the school between 1874 and 1876 while Betti was serving as secretary general in the ministry of public instruction. In 1900 Dini was formally appointed director of the Scuola Normale, a position he held until his death in 1918.

Ulisse Dini was himself a star of nineteenth-century Italian mathematics. Born and raised in Pisa, he had graduated from its university in 1864 and, with the exception of one year of postgraduate study in Paris, he spent his entire career in his hometown. His devotion to mathematics and his students was matched only by his involvement in local and national politics—including a seat on the Pisa City Council, election to Parliament, and membership on the Italian government's advisory board on matters relating to education, curriculum and academic appointments. His appointment in 1871 to Pisa's chair of analysis and higher geometry, formerly held by his teacher Betti, came on the heels of his election to the Pisa City Council. By the end of the decade his role in placing the field of analysis on a solid foundation would secure his place among the titans of nineteenth-century Italian mathematics. In the preface to his treatise "Foundations for the Theory of Functions of Real Variables," which became a classic in its field, Dini recounted how he became interested in the theory of functions of a real variable. "About twelve or thirteen years ago," he began,

> there arose in me the doubt that some fundamental principles of analysis did not present themselves in their statements and demonstrations with all the rigor due to mathematics. I was, however, new to the scientific life and finding that no one else had raised such doubts, I convinced myself that they existed only in my mind. I had to learn from some memoirs of [Hermann] Schwarz and [Eduard] Heine around 1870 and 1871 that men far more experienced in science and meriting great esteem had raised even greater doubts, and that a circle of German scientists were themselves aiming to place the principles of algebra and infinitesimal analysis on a more solid basis.[1]

Like Betti, his teacher and later his colleague at the University of Pisa, Dini had few peers in the classroom. In describing for members of the Italian senate what it was like to be on the receiving end of one of his lectures, the distinguished mathematician Vito Volterra remarked, "He was not only a great scientist, but he was also a *maestro* of exceptional effectiveness ... he taught like a young man, with passion, with a vigor truly extraordinary. In the second half of the century, he undertook a very accurate and detailed critique on the foundations of geometry, of analysis, of mechanics. Ulisse Dini stripped the principles of mathematical analysis of all superfluity and so gave its teaching a perfect rigor, as perfect as a human work can be. His classroom teaching had almost a polemical air: he spoke almost as if he had an opponent in front of him. And he continued to behave that way in school well into his last years when he had already triumphed completely: it was his inexhaustibly youthful mind that made him as combative as in the first years."[2]

In later years, Ricci's own lectures at the University of Padua would be remembered for their precision and clarity. "Whoever wrote them down would never have had to edit them," said Tullio Levi-Civita, who took his own degree in mathematics a generation later under Ricci's supervision. While not as animated in the classroom as Dini, Ricci lectured with an intensity and gravitas that he never lost. He was a staunch advocate of proofs, as can be seen in this excerpt from a letter in which he defended his pedagogical style. "I don't deny," he wrote to a colleague,

> that the proofs of the lemmas, which I must prefix to the definitions that provide for the total in the product, present some difficulty when they are presented to students who unfortunately don't take their education seriously. I don't consider this a sufficient reason to abandon [this approach]. First of all, if my own judgment doesn't deceive me, these proofs are beautiful and rank among

> the things that I care about most in my modest scientific output, and it is not insignificant to me that I am encouraged in this view by those whom I consider my best students. Secondly, since the young people who come from the secondary schools have unfortunately had very superficial preparation in this material, I find it necessary to delve deeper into the topic about rules for real numbers, and I think that [the demonstration of proofs] serves my purpose better than any other method so I will not abandon it.[3]

Dini left an indelible mark on Ricci's approach to mathematics. Many years later, in an exchange of letters on number theory with the University of Naples mathematician Alfredo Capelli, Ricci explicitly contrasted their differing philosophies. Capelli, he said, believed that "the nature of the modern scientific method [is] to eliminate all abstract elements so that using the algorithms as simple instruments of calculation, it becomes natural to present and legitimatize their use with convenient rules and definitions."[4] He then explained how he approached the study of mathematics from a very different starting point. "For me, by contrast, and I think I am in good company, analysis must grasp the material before the application, but must also explain it in a manner that strips it of all that is not essential and makes the mathematical concept clearer and free of every extraneous element. To this way of thinking we owe the great progress that in our era has put science on the road to absolute rigor, and rid it of so many questions that delayed its journey and even called its very basis into question." Abstract concepts in mathematics, Ricci maintained, had a valuable role to play, provided they are not "introduced just for the sake of novelty." Rather he maintained, abstractions should be reduced to their fundamental elements—their essence—if they were to be properly taught, applied, and understood.

The following year, Ricci attended Betti's lectures on celestial mechanics and mathematical physics while working on his doctorate under Dini's supervision. In June 1875 Pisa awarded him a doctor's degree in mathematical physics, with full honors, for his thesis on linear differential equations. "In this thesis [which remained unpublished]," he later wrote, "are collected the results of such arguments by Misters [Lazarus] Fuchs, [Wilhelm] Thomé, [Ferdinand Georg] Frobenius, and others"[5]— German mathematicians who made important contributions to the field of linear differential equations.

With the help of a competitive scholarship underwritten by another professor of mathematics at Pisa, Ricci spent another year at the Scuola Normale doing post-graduate work in higher analysis under Dini's watchful eye. In July 1876, he completed a second dissertation, the *abilitazione,* which qualified him to teach mathematics in high school. (Job openings in Italy's national universities were few and hotly contested, and even outstanding graduates almost invariably began their teaching careers in a high-school classroom. For this the *abilitazione* was an essential prerequisite.) That same month, he satisfied the one remaining requirement, namely to present a lecture to the mathematics faculty. He recorded the event in a terse diary entry: "It was a formality that lasted about ten minutes, after which came the vote on my thesis in which I obtained *la lode* [with praise]."[6]

For his *abilitazione,* Ricci expanded his earlier research to include, as he later explained, "linear differential equations in which the coefficients are rational functions of independent variables and used [the coefficients] to extend, as far as possible, a problem of Riemann's" involving hypergeometric functions. In a brief account of his student's three years, Betti, writing on behalf of Ricci's thesis committee, reiterated its praise for his academic work and noted that the committee "had expressed the opinion that when Dr. Ricci has completed the research that forms the subject of the thesis, this could be inserted in the *Annali della Scuola.*"[7] Here, however, Ricci proved to have a mind of his own. Rather than submit either thesis to the school's journal, an internal publication intended mainly to showcase the research of its own students, he published a revised compilation of both as a single article in the *Giornale di Matematiche,*[8] cofounded and edited by the Neapolitan professor of higher geometry, Giuseppe Battaglini. Established with the modest goal of meeting the needs of students in Italian universities, in time the *Giornale* would become one of the main vehicles for the spread of non-Euclidean geometries in Italy.

Ricci spent 1877 in Pisa, methodically attending both Dini and Betti's university lectures while independently studying a selection of original papers on abelian functions by Neumann, Riemann, and Weber, and reading Maxwell's writings on electricity. In a narrative that he wrote several years later about his scientific activities during this period, Ricci singled out as particularly influential two papers of a mathematical-physical character published in *Nuovo Cimento,*[9] Italy's sole physics journal. They may have been suggested to him by Betti, who from the mid-1860s on worked exclusively on topics dealing directly or indirectly with mathematical physics and mechanics.

It would be wrong to suggest that Ricci's time in Pisa consisted of all work and no play. His brother visited him; he went with friends to the theater; and he most certainly took regular walks along the embankments of the Arno River on weekdays before dinner and on Sundays before lunch. All of these extracurricular activities are mentioned in an extant fragment of his diary, along with a more poignant tale. In passages that run to about two handwritten pages, he tells the story of his unrequited and apparently unarticulated affection for a pretty, fair-haired girl. The entry is not dated. He begins as follows: "At the end of the first year in Pisa, a slender and delicate young girl, unattainable, with blond hair, large blue eyes, and an upright carriage, struck my fancy. I saw her often along the promenade along the banks of the Arno with her father and a friend or accompanied by the mother, and not a day passed that I didn't anxiously search [for her] among the crowd, and seeing her gave me an indefinable sense of pleasure."

He continued his promenades, hoping to catch more glimpses of her. By the end of that year, he had learned her name and that she was being wooed by a fellow student. As the months passed and he pursued his studies, he assured himself that what he had felt was only "simple affection." That is, until one spring evening when he unexpectedly came face to face with her during the intermission of a concert he was attending with a friend. "I turned pale, then I blushed, so much that my friend took stock of my troubled expression and immediately guessed the cause. That night I cut off talk quickly; another night, however, we met again ... we spoke for a bit, and all of this contributed to the discovery of tenderness in my heart that I didn't know I had." For the next two nights, he hardly slept. He encountered the young woman, whom he only ever refers to by the initials L.B., several more times, and confided to his diary that he even deceived himself briefly into thinking he could take the place of her suitor. On the evening of July 5, 1876, two days before he defended his thesis, his fantasy was shattered once and for all. Hurrying to hail a cab for a friend, he passed her home and unexpectedly saw her "there [by the open window] in front of me." Emboldened, he looked straight at her, and from the look she returned at that moment, he resolved never to think about her again. He consoled himself with the knowledge that he would soon be leaving Pisa (although he ultimately returned for another year) and with the thought that "the wound if not perfectly mended, is almost healed." Still, he wrote wistfully, "I would have been happy to be embraced by a pure and burning love, such as the one I felt capable of giving her."[10] Purity and passion, he had discovered, need not be confined to mathematics.

CHAPTER 3

# Munich

As early as 1876, Ricci had discovered that a freshly anointed license to teach mathematics did not pay the rent or automatically open any classroom doors. That year Eugenio Beltrami, one of his examiners at Bologna, left Rome for the chair of mathematical physics at the University of Pavia, leaving his chair of rational mechanics vacant pending a *concorso*—a national competition overseen by the Italian ministry of public instruction—to select his successor. The newly minted teacher sensed an opportunity and entered the competition. Had he consulted professors Betti or Dini beforehand, they might have advised him to lower his expectations and look elsewhere. Nevertheless, in an effort to qualify Ricci for preliminary consideration, Betti wrote the strongest letter of recommendation he could, given that Ricci had yet to publish a single paper. In his report to the ministry, the director of the Scuola Normale described his former student as smart, serious, and earnest. "In his studies he has always distinguished himself as a young person of great intelligence ... [and] in addition to being very talented, Ricci is a deep thinker, good-natured, and has the qualities of an outstanding teacher,"[1] he wrote. Having observed Ricci drilling pre-college students in projective geometry exercises, Betti knew Ricci to be a gifted instructor. He also volunteered Ricci's address in Lugo, thinking officials might offer him a high-school teaching job there. Nothing came of his efforts, or Ricci's. A cursory, undated statement, scribbled in red pencil in his personnel file by an anonymous ministry official, sums up the vetting panel's verdict: "He was not declared eligible."[2]

Two years later, in February 1878, Ricci's mother died after a long illness. She was fifty one. Her loss, he later wrote, was wrenching and spelled "the close of the most beautiful era,"[3] he had known, when life itself seemed indestructible, free of sorrow, bad luck, and other misfortunes. "It is now 40 days since mother died," Domenico wrote to his younger brother in Pisa, "and I still can't get used to such a loss."[4] It may have been partly to assuage his own grief that Ricci threw himself with renewed determination into a new round of academic applications. That spring, now with publications to his credit, he applied for a teaching position at the Polytechnic Institute of Milan. This time he was accepted into the *concorso* and proceeded to assemble the requisite volume of paperwork for the selection committee. His dossier, submitted to the ministry of public instruction, consisted of six certificates, including one from the mayor of Lugo attesting to his exemplary moral character; various transcripts; copies of his doctoral and *abilitazione* theses; and copies of his first three published articles, from the *Giornale di Matematiche* and *Nuovo Cimento*. This compilation was sufficiently impressive to place him second, with a score of 33/50. Later that spring he entered the *concorso* for the chair of algebra and analytical geometry at Bologna and placed fourth.

There is no evidence that Ricci ever considered teaching high-school mathematics. However, the opportunity to broaden his horizons while honing his mathematical skills did appeal to him. When the ministry of public instruction announced in May 1878 a *concorso* for two postgraduate fellowships for study abroad in advanced mathematics, Ricci applied for and received one. The ministry notified him that October that the fellowship, which carried an annual stipend of 3,000 lire, covered the 1878-1879 academic year. On Betti's advice, Ricci chose to spend the year in Munich, Germany, doing advanced work with Felix Klein, whose youthful contributions to non-Euclidean geometry and the links between geometry and group theory had established his reputation and made him a celebrated mathematician.

In his teens, Klein had his heart set on becoming a physicist. But while a student at the University of Bonn, from 1865 to 1868, he had become laboratory assistant to Julius Plücker, who specialized in analytic geometry and who held a joint chair in mathematics and experimental physics. Earlier in his career Plücker had set geometry aside in favor of research on magnetism and spectroscopy, but around the time Klein became his assistant nearly twenty years later, he had reversed course and returned to geometry. He died in 1868, leaving unfinished the second half of his book *New Geometry of Space,*[5] which Klein completed, having become deeply grounded in the subject himself.

In the course of these editorial labors, Klein met the Göttingen algebraic geometer Alfred Clebsch, who was then involved in founding (together with Carl Neumann) the *Mathematische Annalen,* which quickly became the preferred publishing vehicle for work in invariant theory. Clebsch took an instant liking to Klein, brought him to Göttingen (Klein served as a lecturer there in 1871), and actively backed his appointment as full professor at the University of Erlangen in 1872. Clebsch died that same year of diphtheria, leaving Klein to step in as managing editor of the journal. Under Klein's leadership, the *Annalen* became the preeminent mathematical journal in the world, a position it would hold for many decades. Its fame owed much, historian of mathematics David Rowe has written, to Klein's "knack for promoting the works of mathematicians who either were estranged from or stood outside the mainstream influence of the Berlin school [of mathematicians]."[6]

At Erlangen in 1872, Klein delivered the customary address at the formal opening of the academic year. As part of the inauguration program, he had also prepared a new work for publication,[7] *A Comparative Review of Recent*

*Researches in Geometry*, now known simply as the Erlangen Program, in which he illustrated the importance of the theory of groups for the classification of different geometries, including the projective, affine, and metric geometries that had arisen during the nineteenth century. These geometries, he wrote, "can tersely be characterized by specifying the groups of transformations that leave invariant the relevant relations."[8] Deeply intrigued by the unifying prospects inherent in the group idea, Klein worked for many years in elaborating, using, and promoting the concept. The influence of his Erlangen Program can be felt to this day in many college-level geometry textbooks.

In 1875, Klein left the University of Erlangen and accepted a chair in mathematics at the Polytechnic School in Munich.[9] Alexander Brill, a former student of Clebsch's, was appointed to a second chair for advanced mathematics at the school as well. With its lavishly decorated palaces and monumental buildings, numerous museums, art and sculpture galleries, and theaters partial to the music of Richard Wagner, Munich had grown into a modern city during the nineteenth century. Its Polytechnic School, founded in 1868 by King Ludwig II of Bavaria in response to the industrial expansion under way in southern Germany, provided a new generation of young engineers with the sophisticated mathematics and equipment they needed to solve a wide range of novel technological problems, from designing machinery to building electric power stations. In addition, Klein had arranged for a set of rooms to be set aside as a mathematical institute. There was also a studio for the production of three-dimensional geometric figures in plaster of Paris, space in which to exhibit them, and funds and an assistant to support this work. He also encouraged colleagues elsewhere to build their own mathematical models. In a letter to the eminent Italian geometer Luigi Cremona, he wrote, "It has always seemed to me that for geometrical problems not only are the theorems important... but also important is the direct visualization of these objects."[10]

Ricci arrived in Munich in November 1878, in the middle of Munich's winter semester,[11] found lodging near the school, and met with Klein. After a few pleasantries that included sharing a letter of introduction from Betti, Klein proposed what he considered an eminently reasonable course of study. Ricci found the list of assignments daunting but concluded they would be doable—with a little tweaking on his part. As the dutiful student wrote to "Chiarissimo Signor Professore" Betti the following month, Klein "believes only two [courses] could be useful for me, his on the theory of [algebraic] equations and Professor Brill's on elliptical functions." To derive the greatest benefit from the lectures, added Ricci, he needed, in Klein's opinion, firsthand knowledge of "several theories," and he had, on Klein's advice, begun to study the French mathematician Joseph Serret's papers on number theory and the section in his advanced algebra textbook on the theory of substitutions. "Moreover," continued Ricci, Klein wanted him to delve into the second part of Clebsch's course of geometry published by Ferdinand von Lindemann. "Added to this I need to bring myself up to date with the lectures Klein gave before my arrival, and you will understand that I am not lacking for work, nor do I need any right now."[12]

Within a few weeks of his arrival in Munich, Ricci had stopped going to Brill's lectures. As he explained in detail in the same letter to Betti,

> I am not very sorry, because I had already done the theory of elliptical functions [in Pisa] under Professor Dini. In this way I have been able to turn all of my attention to Klein's extremely interesting course and to the various theories . . . several of which, especially the geometrical ones, are entirely new to me. I even see that it would be very useful to me to be able to study Riemann's paper on abelian functions, and I will do it if in the future I find time, which I lack now. Meanwhile, I looked at the first two chapters on the theory of numbers in Serret and the first four [chapters dealing with] substitutions, arguments I feel I would find much pleasure in studying because it seems to me there is still much to do, such that I am looking forward to dedicating myself more to them with greater calm and purpose.[13]

In class, he reported, Klein had been lecturing about the solution of equations in general, including a discussion of the Tschirnhaus transformation in algebra.

Aside from his crushing workload, Ricci declared himself to be "exceedingly well" and, in closing, thanked Betti, once again, for having made it possible for him to make "the acquaintance of a person as accomplished and kind as Professor Klein, from whom," Ricci declared, "I have the most vigorous help in my studies."

By mid-March 1879, which marked the end of his first semester in Munich, Ricci had progressed much farther in his independent studies, judging from the semiannual report he wrote for the ministry of public instruction. In it, he described his efforts to catch up with Klein's lectures, beginning with Serret's *Corso d'Algebra Superiore*, before turning to a work by Paul Gordon dealing with groups of substitutions of algebraic equations. The Italian postdoc had also been introduced to the professor's own research on equations of degree greater than 4. As Ricci noted, "I am also engaged in and still occupying myself with several questions put to me by Professor Klein, which go back to what the same professor treats in his most recent papers, 'On the equations of the icosahedron' and 'On the transformations of degree 7,' [and] I am [also delving into] the 'Theory of Linear differential equations,' which in the past were the subject of several of my studies."[14]

In his own first report to ministry officials on Ricci's academic performance, Klein recounted how Ricci met with him frequently, followed "with impeccable regularity" his lectures, which were open to advanced students, and took an active role in Klein's mathematics seminar, where the participants typically reported on new journal articles or on their own work. Ricci's knowledge of mathematics, Klein went on to say, had benefited from his wide reading in the literature and through proofs he had solved partially by himself. "I can summarize my opinion about him [Ricci], affirming that he has profited by his stay here in taking on a lot of work and at the same time seeing results, and I confer this certificate [signifying completion of the first semester] voluntarily, the more so since I participate fully in his mathematical development," Klein declared.[15]

Ricci filed his last report, dated August 1879, to Roman officials from Lugo, just after the end of his second semester's work in Munich. Much to his delight, Klein had offered the theory of elliptic modular functions—in particular transformations of degree 5, 7, and 11—as a special course for advanced students that semester. "I set myself the task of gaining a clear insight into this whole domain with the aid of geometric function theory," Klein later wrote[16] of this work, which overlapped with Ricci's stay in Munich. This new course, Ricci explained in his report, offered "the double advantage of presenting to me an extremely important application of the theory of Galois on algebraic equations, which I had studied in the previous semester, and of hearing from the same mouth of their illustrious inventor the working out of new methods that attract the attention of those in Germany and outside who are concerned with this most important theory of elliptic functions."[17] During Munich's summer semester, the aspiring mathematician gave a talk in Klein's mathematics seminar on a paper on modular equations by Henry John Stephen Smith, professor of geometry at Oxford. Ricci's personnel file in the ministry of public instruction includes a second certificate by Klein testifying to his performance in Klein's course on theories regarding modular functions and their elliptic transformations. In it, he wrote simply "I can affirm that he has fully understood the significance of the theories treated by me."[18] With Klein's affidavit in his pocket, Ricci returned to Pisa that fall.

In many ways, Ricci came of age as a mathematician during his year in Munich. Perhaps most importantly, he acquired a healthy measure of self-confidence that would later sustain him in the face of repeated professional and academic setbacks. In his approach to educating aspiring mathematicians, including Ricci, Klein seems to have followed a formula handed down by his own teacher. "The most important aspect of Clebsch's influence was the moral influence he exerted by instilling in us, in addition to a deep interest in science, a confidence in our own powers,"[19] Klein once wrote. The English mathematician Grace Chisholm Young, who took her doctorate at Göttingen with Klein many years later, remembered him in a similar light, as a mathematician with a "rare type of mind... teem[ing] with ideas and brilliant reflections." In class, she added, "his favorite maxim was 'Never be dull!'"[20] Of the many words later used to describe Ricci's calculus ("primitive and complex analytical apparatus"; "algorithms... developed are useful but not indispensable"), none ever implied that his powers of invention were lacking. As for Ricci, the only known evidence of his own feelings toward Klein appears in the signature line of a letter that he wrote to his former mentor many years after he left Munich. It reads: "Your devoted disciple, Gregorio Ricci."[21]

CHAPTER 4

# Padua

In early November 1879, shortly before the term began at Pisa, Ricci sought out his former calculus teacher, Ulisse Dini. Eager to plant his feet on the academic ladder, Ricci wanted to know if there were any postdoctoral positions within the university suitable for a young graduate with his background. Ricci's question resonated deeply with Dini. His colleagues in Germany relied on voluntary assistants, while in Italy the state covered the expenses for the recently established position of university assistant. Keen to boost his student's career prospects and perhaps also to put Pisa on a level playing field with other Italian universities, Dini paid a visit to the capital, where he had a private chat with a well-placed official in the ministry of public instruction. Upon his return to Pisa, the politically astute professor of mathematics reached for pen and paper.

Dini's petition, addressed to the university's rector, opened with a detailed description of his own teaching duties and other responsibilities. "I have expanded my infinitesimal calculus lectures in the past several years," he began, and that, "together with a thousand other jobs that descend on me from the higher analysis [course], and a book in press on Analysis, have left me with an onerous, almost impossible burden, I would say, which affects the mathematical drills." He added that, as much as he would like, he did not have the time to help "the young men to solve some problems by applying the theories that I develop in [my] calculus lectures."[1]

Greatly concerned that his students should derive the greatest possible benefit from his lectures, Dini pointedly noted that instructors in other state-supported institutions of higher learning in Italy, particularly those teaching infinitesimal calculus, had their own assistants. In the course of requesting his own assistant for the 1879-1880 academic year (to be paid, naturally, at the same rate as assistants elsewhere), Dini nominated Ricci for the position. His former pupil, he assured the head of the university, was "one of [our] most distinguished students." He had "obtained eligibility in a concorso for a university chair," and perhaps most importantly from Dini's point of view was ready to start at once.

The following day Dini dispatched a short note to his inside contact in Rome, assuring him that Pisa's rector had immediately contacted ministry officials there about Dini's need for an assistant. Along with asking his unidentified correspondent if he wouldn't mind doing a little lobbying on his behalf, Dini volunteered that the preferred candidate for the position, "Dr.

Gregorio Ricci Curbastro of Lugo is a remarkable young man in every respect."[2] Within a week Ricci had been appointed Dini's temporary assistant at a monthly salary of 100 lire.

At the same time, Ricci continued his earnest quest for a permanent professorial position. By mid-February 1880, he had filed applications for vacant chairs of mathematical physics at Rome, Naples, Bologna, Padua, and Palermo and assembled the portfolio of his papers, certificates, and other affidavits to present to the ministry of public instruction. In his spare time, he busied himself reading the German mathematician Eduard Heine's treatise on spherical harmonics.

Ricci also petitioned ministry officials for the title of *libera docenza* in differential and integral calculus at Pisa. The prerequisite to being put in charge of a course, the designation also conferred the right to teach students for a fee. The examining commission—composed of Dini, Riccardo Felici, and Ernesto Padova—unanimously agreed after reviewing Ricci's published papers that he possessed the necessary qualifications to offer private lessons in calculus. That same month, however, Ricci learned that he had been hired as an associate professor of mathematical physics at the University of Padua, effective December 1880. The need for an interim appointment faded.

Situated a dozen miles inland from Venice on the Adriatic coast, the University of Padua was the second oldest university in Italy, having been founded in 1222 by a group of professors and students protesting conditions at the University of Bologna. Through tumultuous centuries it had served as a beacon of tolerance and academic freedom for the generations of students and faculty who occupied a building called *Il Bò*, a grand palazzo that took its name from the *Hospitium bovis* (formerly a hotel with a sign of an ox). Between the rise of the free republic of Venice in 1405 and its fall in 1797, the university constructed a permanent anatomical theatre (where it pioneered public dissections and autopsies), hired learned professors (among them Andreas Vesalius, Johann Müller Regiomontanus, and Galileo Galilei), attracted a diverse student body (including William Harvey, Nicolaus Copernicus, and Gerolamo Cardano), and founded an astronomical observatory. In 1813, following two brief periods of rule by the French, Austrian troops occupied the Veneto region, including Padua, the capital of the province, for more than fifty years. Shortly after the end of the Third Italian War of Independence in 1866, in which the troops of King Victor Emmanuel II defeated the Austrian army, the Veneto region was annexed to the new Kingdom of Italy by a plebiscite.

Padua's university gained parity with the University of Rome in 1872, following the passage that year of a parliamentary measure that extended the provisions of the Casati Law. These were a comprehensive set of articles governing programs, personnel, and promotions throughout the Italian educational system, including the mathematics faculty that Ricci joined eight years later. In effect, the ministry of public instruction, which typically deferred to its *Consiglio superiore*—an advisory board known as the high council, on such matters—could, and did, play dice with the fortunes of Italian academicians, a situation that would profoundly affect Ricci's career.

Soon after his appointment, Ricci sent Enrico Betti, Pisa's acclaimed author of numerous publications dealing with problems of mathematical physics, an urgent message concerning his first classroom lecture. "I have learned that at Padua," he wrote Betti, "they are in the truly regrettable habit of having the new [associate] professors inaugurate their courses with an opening address attended by many colleagues and pupils in the mathematical, physical, and natural sciences faculties. Many people have told me to talk about something of no importance, but I need to take this seriously, although a mathematics course especially presents difficulties."[3]

Given a choice, continued Ricci, he would prefer to discuss a historical or scientific subject, or even something farther afield, such as the teaching of advanced mathematics in Italy "but for now I don't know where to find the material before getting down to work of this type. Would you be so kind," he continued, "as to shed some light on this [and advise me] on the topic to discuss . . . and as to the books, which ones to consult for references? I don't need to tell you that while I am sorry to take advantage of your kindness, it will only give me more reason to feel grateful to you." In closing, he also reminded Betti of an earlier promise to send him a memo about the essential books in the field of mathematical physics, an offer Ricci apparently considered too important to let slide.

Only an undated draft of Betti's reply, a few hastily scribbled lines on a scrap of paper tucked among his correspondence files in Pisa, has survived earthquakes, two world wars, and the passage of time. "*Caro* Ricci," he began, "It seems to me that a subject for an introductory lecture could be the influence of mathematical physics on the progress of mathematical analysis." This was followed by a scramble of incomplete, sometimes illegible, words and phrases: "The... analytic of functions of the actual quantity... originate from... or from the theory of heat. The theory of functions of complex variables of which Riemann has found the... Gauss and Green... originate from the theory of electrostatics and magnetism."[4] The mathematician Vito Volterra, who overlapped with Ricci during his last year at Pisa, described Betti as someone who lived and breathed theoretical physics and displayed "the greatest enthusiasm and desire to use it in a practical way."[5] It was his usual habit, Volterra added, "to reconcile analytical concepts with natural phenomena." It would have been unthinkable for Betti to encourage Ricci to talk down to his audience or to speak on a topic about which he knew very little.

More than two decades later, when Ricci was invited to present the inaugural address at the formal opening of the 1901 academic year at Padua, he delivered a well-researched, accessible, and impassioned speech about the origins and development of modern concepts in geometry. His goal, he told the assembled faculty and student body, was not merely to give them some idea of the importance and historical development of geometry during the previous century's "golden age" of mathematics, but also to illuminate "the sovereign beauty of mathematical studies and the reasons for its allure, which it exercises on so many and such noble intelligences."[6] Perhaps, Betti, who died in 1892, would have approved, given the solemnity of the occasion and the dozens of faculty, ranging in disciplines from astronomy to zoology, who had robed up according to their rank and gathered in the regal public hall of the university to hear the principal—and only recently promoted to full professor—speaker.

On January 8, 1881, when Ricci opened his mathematical physics class at Padua with the requisite inaugural lecture, he began with some indications about the direction of the course. Afterward he addressed the audience directly, saying, "Students like you understand the immense advantages of these methods, which are especially useful for approaching what has been for a long time the ultimate goal of physics: I am speaking of the unity of science, which is the discovery of the relationships connecting various natural phenomena, and the causes on which they depend."[7]

After reading the complete text of this lecture, which had been sent to him by one of Ricci's grandsons in 1925, his long-time colleague and scientific collaborator Tullio Levi-Civita quoted these lines to open his own commemoration of Ricci, which he delivered in January 1926 at the Accademia Nazionale dei Lincei (literally, the Academy of the Lynx-Eyed), in Rome.

Although he later dramatically altered the arc of his research, Ricci's initial publications at Padua suggest he took to heart the advice Betti had offered him. As transmitted to Ricci by his own teachers, Klein included, the branch of mathematics known as mathematical physics encompassed "the whole domain of 'phenomenological' physics that works with differential equations, as it has been developed by [the Germans and the French] and by the English, culminating in Maxwell's equations, that is, the physics that works with the idea of continuous media."[8] In his first course, Levi-Civita later reported, Ricci developed "the applicability and use of Green's functions in the theory of potentials" and then proceeded to demonstrate the "equivalence of galvanic currents (constant in time) and permanent magnets."[9] Armed with the mathematical formulation of this equivalence, Ricci derived "the explicit transformation of the respective potentials" for the most general case. Two papers—"On the potential function of conductors of constant galvanic current," the paper he published in 1882, and "On several systems of differential equations," which appeared the following year—summarized Ricci's results. Had he continued to research and write papers in this vein—that is, to follow in the footsteps of Beltrami, Betti, and Dini, who were the leaders in this field in Italy at the time (and whose work he cited in the 1882 paper)—it is likely that Ricci would have had an easier time ascending the country's academic ladder.

In mid-January 1881, Ricci's older brother, Domenico, wrote him a congratulatory note from Lugo:

> Dearest Brother,
>
> As you can imagine, I contemplated with the greatest interest all the particulars of your first lecture as a professor in a university of Padua's caliber, and if I am not mistaken, you can be as pleased as I am.... You must have made a good impression on your students as well as your colleagues and all those who contributed to your nomination. [It confirms my] belief that you turned down an assignment that could have paid you more (offering a *libera docenza* preparatory course), but which for you has no advantage, although it would have [benefited] your students. In so many respects you have begun well, and I have no doubt that in the future you will do even better.[10]

Over the course of more than four decades, Ricci added other topics to his teaching repertoire, including infinitesimal analysis and higher geometry. But abandoning his first love, mathematical physics, was not something he ever seriously considered. Even when he finally secured his promotion to full professor in 1890, Ricci continued to teach his original course in mathematical physics.

Accounts of Ricci's classroom performance are scarce. Although he taught at Padua for nearly half a century, no school of mathematics grew up around him. Numerous students came and went, among them Levi-Civita, his future collaborator, who entered the university as a freshman in 1890. Years later he was to recall that although his professor was not a charismatic speaker, his lessons "were admirable for their precision and their extreme fluidity of form: whoever wrote them down would never have had to revise them."[11] The Italian analyst Angelo Tonolo (1885–1962), who took a degree in mathematics at Padua in 1908 and went on to spend his entire career there, also considered Ricci his mentor; indeed, he devoted a good portion of his scientific career to solving a host of problems in classical mechanics and differential geometry, using the methods of Ricci's calculus. In a talk given on the occasion of Ricci's centenary in 1954, Tonolo recalled his professor's lectures as "a model of precision, never any hesitation,"[12] delivered by a tall, dark-haired and well-mannered scholar in a gentle, mild voice. "The serious expression" he added, "illuminated by intelligent and soft eyes, was tempered sometimes by a friendly smile." Modest in demeanor, imperturbable, and exuding personal dignity, Ricci impressed him as "the dignified figure in compassionate contemplation, the sober gesture, and the carefully weighed and calm word, which gave him a diffident and distracted air. But those who were willing to look beyond this aura of reserve would have found a refined, receptive human sensibility, warm friendliness, and the greatest tenderness for the family."[13] As befits "the true passion of a disciple," Tonolo is credited with being the driving force behind the publication of Ricci's works in the 1950s, "coordinating and supervising the work of various collaborators,"[14] a colleague later noted.

Ricci's arrival in Padua coincided with a major shakeup in the university's mathematics faculty; within a year or so, two-thirds of the professors had been replaced by new faces. These included Giuseppe Veronese, an algebraic geometer who succeeded Giusto Bellavitis, inventor of the theory of equipollents, whose death in 1880 created a vacancy; Ernesto Padova, who relinquished his professorship of rational mechanics at Pisa in favor of the chair of advanced mechanics at Padua; and Giovanni Garbieri, an associate professor, who taught complementary algebra there. Francesco D'Arcais, who had joined the faculty in 1878, covered infinitesimal and higher analysis, while the most senior member of the mathematics staff, Domenico Turazza, an expert in hydraulics and the founder and director of the school of engineering, was in charge of rational mechanics and descriptive geometry. Giuseppe Lorenzoni, who taught astronomy and geodesy, was also a member of the science faculty.

Padua's teacher-training institute (Scuola di Magistero), which the university had established within its science faculty a few years earlier, took a keen interest in these new appointments. Chartered by the minister of public instruction in 1875, with the aim of creating a professionally trained corps of teachers in the newly unified nation, these pioneering university-based institutes had a rocky beginning, in part because the legislation itself "was fundamentally ambiguous, emphasizing both research and professional teacher training," according to the mathematics historian Livia Giacardi.[15] Before then, aspiring secondary-school teachers did not need a degree to teach. At Padua, the faculty of science had proceeded to organize courses in chemistry, physics, and natural sciences for upperclassmen planning to become secondary-school teachers, but ministry officials in Rome had blocked repeated requests by the university to create a *Scuola di Magistero* of mathematics, citing the lack of qualified instructors. In fall 1882, matters came to a head, after a ministry promise to have Veronese teach the required higher geometry course at university expense fell through for budgetary reasons. The science faculty, which found it difficult to believe that the treasury did not have 1,200 lire to pay Veronese for the one extra course, nevertheless persuaded their colleague to offer it as a class whose cost would be underwritten by the students, provided the ministry accepted that solution. While waiting for a response from Rome, a sizable number of third-year mathematics and physics students petitioned the rector to intercede with the minister on their behalf, citing a "common desire to learn more by attending the lectures prescribed for this school and our great eagerness to obtain at the end of these studies the diploma allowing us to teach mathematics in secondary schools."[16] Once again, the high council turned down both requests, defending its decision on the grounds that many of the lessons would be entrusted to new professors or to faculty who were "already committed to teaching... two of the most important subjects in the last two years of instruction, higher analysis and higher geometry"[17] for the doctor's degree in mathematics.

While the science faculty continued to bemoan the lack of a mathematics section within the Scuola di Magistero, Ricci encountered no difficulty in giving a series of lectures there for students studying for the doctor's degree in physics, in addition to teaching mathematics to his own students. Between winter 1882 and summer 1883, he met regularly with twenty students for a total of twenty-two lectures. The following July, eight students obtained the *laurea* diploma in mathematics, but in the absence of an instructor for higher geometry were not able to obtain the crucial *abilitazione* diploma that would have certified them to teach. Once again, the rector passed along to Roman officials a resolution from his science faculty protesting the ministry's delaying tactics.

This time, however, a combination of events—Veronese's sudden appointment as an instructor for higher geometry, a change in the leadership of the ministry's high council, and an 1884 article in one of Padua's newspapers—brought a different resolution to the long-standing dispute. Citing anonymous sources who said that a mathematics section had become a fait accompli, now that Veronese's lectures had begun, the article went on to report that the ministry had still not officially notified the Scuola director, and that there were regulations laid down in Rome that needed to be followed. But this time, the high council saw matters in a different light. As Francesco Brioschi, the council's vice-president and a celebrated mathematician in his own right, informed the minister, "Taking into account the distinct abilities overall of the professors who represent the mathematical sciences at the University of Padua, the mathematical section of this council has given notice that you can establish the requested Scuola di Magistero for pure mathematics."[18]

Several years later, Veronese would hand over the teaching of higher geometry in the Scuola to Ricci. The lectures were scheduled for Saturday afternoons, and the earliest classes gave Ricci's students a foretaste of their professor's "diffident and distracted" aspect. As one of them later recalled, "After a couple of weeks, waiting in vain for the professor to appear, my classmates and I were leaving... when he arrived in a carriage, completely out of breath (evidently he had [walked] part of the way), ran through the door, and went into the lecture hall, followed by us. Seating himself in front of the blackboard, with his eyes lowered like a guilty schoolboy, he confessed, 'I forgot. This will not happen again.'"[19]

CHAPTER 5

# Math and Marriage

In the winter of 1884, on the eve of his thirty-first birthday, Ricci began courting a young woman by the name of Bianca Bianchi Azzarani, who lived in nearby Imola, a small town dating back to the time of Cicero. Shy and reserved, and perhaps also fearing the disapproval of his father, who had many harsh words to say about the woman his older brother, Domenico, had fallen in love with, the wife-seeking mathematician had enlisted the help of his parish priest in selecting a suitable mate.

Domenico's quarrel with his father had taken place while Gregorio was a postgraduate student in Munich in 1879, and Antonio's side of the story reached him soon afterward. Not only did the young woman in question lack a proper dowry, according to the letter Gregorio received from his father, she also came from an unconventional family (the men and women ate in separate rooms; the mother on occasion ran after her son brandishing a knife). Worse still, she belonged to a branch of one of the numerous Ricci Curbastro families living in Lugo. "Do you think that God can bless marriages between close blood relatives?" Antonio asked Gregorio.[1] How could a Christian youth raised in a God-fearing family like Domenico's abandon the Roman Catholic teachings that his staunchly observant father had worked so hard to instill in him? "And don't we in Lugo, have Imolesi, Faentine, Ravennate—women who are the very models of brides and mothers?" his father demanded of Gregorio. Coming to the heart of the matter, he added, "And have not such marriages been arranged by wise and kindly relatives and friends. . . rather than. . . through the choice of bride from the outskirts of a small village, and worse still, from a family of relatives?" Ostensibly directed at Domenico, Antonio's displeasure may also have served as a warning shot across the bow for his younger son, still an eligible bachelor in 1879.

In any event, once Gregorio made up his mind, several years later, to marry, he opted for a matchmaker, putting his domestic future in the hands of the family's parish priest in Lugo. This priest in turn was friendly with the local priest at the cathedral of S. Cassiano in Imola, where the worshipers included Bianca and her mother. There is no record of whether Gregorio expressed any hopes or preferences regarding a wife; in fact, it may well have been his father who contacted the priest on behalf of his son. Having identified a suitable candidate within the congregation and secured the approval of the prospective bride and respective parents, the clerics duly arranged for Gregorio, who typically returned to Lugo from Padua during the university's winter recess, to meet Bianca that December in her home in Imola.

An only child who had recently lost her father, Bianca was twenty-four, raised in a traditional, well-to-do Catholic family, firm in her convictions and religious beliefs, and ready to wed. Vivacious, cultivated, and intelligent, she came of age at a time when some Italian universities had opened their doors to female students, but her expectations seem to have been the conventional ones for young women of her class and era—marriage, children, and management of a household, including servants. Like most young Italian women of her day, she was probably schooled at home, and while her letters to Ricci make it clear that she enjoyed reading, they offer no hints that she had spent any extended time outside the ancient village of Imola, population 11,000. An oil painting of Bianca as a young woman shows a petite, well-proportioned *Signorina,* with a fair complexion, finely chiseled nose, soft brown eyes and abundant wavy brown hair swept back from a broad forehead. The photographer who took a picture of Ricci surrounded by a group of students several years before he met Bianca captured a fairly tall, intense-looking young man with a full head of hair, a well-trimmed beard, and generous mustache on his upper lip, peering intently at the camera. The first meeting between these attractive, expectant young people must have gone well, because shortly afterward, Ricci wrote to Bianca, expressing his own feelings about marriage and inviting Bianca to share hers.

"Dear Professor" (*Egregio Professore*), she replied on January 5, evidently pleased by the receipt of his letter, but not yet sufficiently emboldened to address him by his first name. "I cannot deny how the kindness of your letter has given me a joyful surprise, noticing from it that your way of thinking and feeling agrees with and is similar to my own." Concerned that she might have made a poor first impression, Bianca explained that she had been overcome by awe in meeting him face to face. "I probably appeared excessively cold, timid, and perhaps also awkward,"[2] she wrote apologetically.

Clearly, although neither party had said anything explicit, the prospect of a proposal was in the air; and in that same letter Bianca did not hesitate to suggest that she and Gregorio appeared to be well matched. "I know and appreciate the value of your just and sensible reflections; and I share the desire you expressed of visiting me as often as you wish when you are in this area, so that we can come to understand each other more and assure ourselves of the reciprocity of our affections, thoughts, and feelings," Bianca wrote to her suitor. And despite her meek demeanor at their first meeting, when it came to putting down on paper her thoughts about marriage, the words flowed. "It is within the domestic walls," she proclaimed,

> among the pure and holy joys that family provides, that a man might find a wholesome and beneficial balm to

> counteract the disillusions that most of us meet in society, which is often grim and evil. Without such a source of comfort, a man with a kind and loving heart would be very sad indeed. Likewise, the woman who finds in her partner a friend, a support, a confidante with whom to share even the most hidden thoughts will always be able to challenge, with a joyful and confident heart, the sufferings that unfortunately are an integral part of the human existence.

Speaking of her own parents, "who loved each other with the same affection till the last day that God willed them together," Bianca expressed her hopes that her "future partner" would be someone like her late, "saintly and virtuous father." Nor was she reticent about urging Ricci to return to Imola sooner rather than later. "If tomorrow, you are thinking of visiting me," she pointedly remarked, "I would be very pleased."

A man on a mission, Ricci took the train from Lugo to Imola several more times during the one remaining week of his Christmas break. When the winter weather permitted, the couple took long daily walks around the outskirts of the town; their evenings they spent indoors, exchanging "pleasant conversations and affectionate expressions,"[3] seated together on the sofa. By the time Ricci boarded the train for Padua on January 13 to resume his university duties, and barely two weeks after first setting eyes on each other, they were ready to announce their engagement. In a letter written a week later to "dear Gregorio," Bianca reminded him, "Remember, I am counting on seeing you again in early February, as you promised me before leaving; I cannot reconcile myself to the terrible idea of waiting until Lent."

In another early letter, plainly reveling in her new status, she related how her friends had reacted to her engagement with good-natured teasing, telling her "that I couldn't have happened to a ... *more unsuitable person* than you ... my good Gregorio."[4] Between parrying mock compliments and shooing away people who wanted to meet her intended ("the brief moments you will be in Imola are too precious to set aside to others"), Bianca, newly affianced to one of the very few accredited Italian teachers to have advanced to a prestigious university position, had suddenly become the talk of Imola. As she informed her fiancé, "With the exception of the closest relatives, those who want to enjoy becoming acquainted with you will have to satisfy their curiosity when they see us [walking together] outdoors." Certainly she saw herself in a new light: "I'm a new object now in Imola," she declared rather proudly, "and with the knowledge that I am now a newly engaged woman, it's as if I had acquired an exceptional personality."

An enthusiastic correspondent, Bianca kept up a steady flow of letters to her "dear Gregorio" during their six-month engagement. Although it is clear from the content of her letters that Ricci also wrote regularly, only Bianca's side of the correspondence, fifty-eight letters in all, written between

January 8 and August 23, 1884, two days before their wedding in Imola, has survived. Did she destroy his letters or perhaps leave them for her family to deal with after her death in 1914? Or did they simply vanish during the chaos that two world wars brought to northern Italy? We can only guess. "If it were possible," she confessed in one of her first letters, "I would correspond with you—I want you to know—not every day, but every hour or minute."[5] And now that they were engaged, she added, did he recognize "the woman he had only recently called the *gentilissima Signorina,* who persistently kept her eyes to the ground ... not daring to raise them to the serious and straight-faced *Egregio Professore* standing in front of her?" Recalling her early shyness "to the point that at times I was embarrassed to talk to you," she teasingly thanked him for his role in her transformation, allowing that "by opening [his] mouth here and there to a small smile and gradually leaving behind that slight hint of condescension," he had gradually led her to overcome her initial state of mind. "Today, [the shyness] has completely disappeared, and has left no trace."

Judging from her responses to them, Ricci seems to have peppered his own letters to Bianca with advice. To overcome her self-described habitual indolence, he counseled taking walks (not her favorite form of recreation, apparently); to counteract stomach problems, he suggested a glass or two of wine (which she forgot to drink); when she proposed buying (with his money) highly sought-after artisan lace from the Lombard city of Cantù for their future home, he assured her that his family would furnish their house with luxurious linens.

They had in common a love of music and smoking. Bianca practiced the piano diligently every day, and during her engagement wrote to Ricci about the new musical scores she was studying. They shared a liking for opera, and when Bianca learned that Amilcare Ponchielli's opera *La Gioconda* would open in Lugo that fall, she made a mental note that the two of them should see it. Ricci enjoyed Cavour cigars, no longer manufactured, but reputed to have had a strong and bitter taste; Bianca smoked cigarettes, proudly putting herself in the company, as she put it, of other "dissipated"[6] women. As the husband-to-be sat in the smoking compartment of the train bearing him back to Padua from Imola in early February, he started thinking about the first weeks of married life: How would the newlyweds manage to come up with a meal in their new quarters on their first day in Padua together? he wondered. "You make me laugh," his future bride wrote back, assuring him that a kindly saint would surely deliver them from such pressing problems and, besides, she could deal with the situation, if necessary.

With his wedding a few months away, Ricci's life that spring revolved around his teaching duties, finding a domestic domicile in Padua, and starting the construction of his absolute differential calculus, which he methodically elaborated in a series of papers published over the course of a decade, from 1884 to 1894. These ranged from ten-page notes in the *Annali di*

*Matematica pura e applicata* to the first systematic presentation of his calculus in a publication issued in connection with the 800th anniversary of the University of Bologna. Ricci's first paper on this topic, completed in Padua in February 1884, dealt with the theory of quadratic differential forms and their transformation properties (these are now called "covariant 2-tensors"). The research problems he had earlier worked on under Betti and Dini's guidance had been consigned to a desk drawer and replaced by investigations into a challenging new branch of mathematics involving analysis and differential geometry.

Building on the fundamental work of Gauss and Riemann on the foundations of geometry in the first half of the nineteenth-century, Ricci's first paper marked his entrance into this new mathematical arena where the rules of non-Euclidean geometry held sway. In 1827 Carl Friedrich Gauss, considered one of the greatest mathematicians of all time, had published an important treatise covering his investigations of curved surfaces in two dimensions. Gauss had also arrived at a theory of a non-Euclidean geometry after failing to prove Euclid's parallel axiom; but fearing, as he told friends, the derision and controversy that would greet such a new and perplexing geometry, he did not publish it. Less fearful of the consequences of such chatter, Russian mathematician Nikolai Ivanovich Lobachevsky and his Hungarian counterpart Janos Bolyai independently constructed and published a non-Euclidean geometry in 1829 and 1832 respectively. Their discoveries brought them neither fame nor followers. Most mathematicians of that era simply continued to work on projective and algebraic geometries.

Two decades later, however, Bernhard Riemann, a mathematician Felix Klein once described "as the only proper pupil of Gauss, the one who has penetrated into his deepest ideas,"[7] was quick to grasp this work's importance. In 1854, in his probationary lecture[8] before the philosophical faculty at Göttingen, which Gauss, then seventy-seven and infirm attended, Riemann offered a lucid and penetrating discussion dealing with the hypotheses underlying geometry's foundations. Mindful of his audience perhaps, the lecture contained no calculations, and was only published in 1868, two years after Riemann's death. In his talk, Riemann advanced an all-encompassing view of geometry, in which space, which he now referred to as a manifold, consisted of any number of dimensions in any kind of space. The distance between two points that are infinitesimally close together on a curved surface in such a manifold (now known as the Riemannian metric) was defined by way of a positive quadratic differential form. Six years later, in 1861, in a memoir submitted to the French Academy of Science, which was offering a prize for the best essay on the distribution of heat in solid bodies,[9] Riemann provided a copious quantity of formulas and, in passing, his "remark that some analytic expression could be illustrated as the curvature components of a metric manifold,"[10] according to Olivier Darrigol, a French historian of mathematics, who has carefully studied both the 1854 and 1861 papers.

The significance of this work, in which according to Klein, Riemann "developed the whole apparatus of quadratic differential forms, now used in the theory of relativity,"[11] apparently escaped the Academy's notice. No prize was awarded that year; and Riemann's entry languished in the archives of the French Academy until 1876 when it was published in his collected works.

After the posthumous publication of these two lectures, Riemannian geometry came into its own as a vibrant branch of mathematics. By the time Ricci had taken up this subject in earnest in 1884, he had mastered the key papers in the field. Indeed, among the manuscript material in the Ricci papers, now deposited in the *Biblioteca Trisi di Lugo,* are four undated pages of calculations by Ricci on Riemann's curved metric spaces, a concept that finally made the theory of general relativity achievable. As Ricci had discovered while devouring this literature, the mathematicians who took Riemann's writings on geometry and quadratic differential forms as their own starting point had carved out different approaches to the problem. Some of Ricci's more senior contemporaries, such as Rudolf Lipschitz and Eugenio Beltrami, preferred the algebraic aspect of Riemannian curvature, seeing similarities drawn from rational mechanics. Geometrical analogies served as a starting point for many of Richard Beez's investigations on the curvature of space; while Elwin Bruno Christoffel aimed to find the conditions of equivalence of two quadratic differential forms.

As a mathematician who prized clarity, precision, and rigor above all else, Ricci felt sure that this branch of mathematics was ripe for improvement. As he wrote in the introduction to his 1884 paper,

> The aim of this writing is to initiate a series of researches on the theory of quadratic differential forms; being founded on purely analytical concepts, these researches will most effectively introduce us to the knowledge of the nature of these forms and will thus avoid all lazy discussions about the existence and nature of spaces of more than three dimensions. Interpretations dictated by geometrical or mechanical analogies—such interpretations can indeed be given—will only be illustrations of such a theory.[12]

Theory, in other words, was the first order of business for Ricci before he would allow himself any discussion of geometrical, mechanical, or physical applications.

Ricci's absorption in this work may have partly accounted for his difficulties in resolving an entirely different preoccupation: making suitable living arrangements for himself and his bride-to-be. Bianca, who took an active interest in the organization of their future home, had urged him to choose an apartment that included a pantry. "This is the first advice that I am giving you as manager [*reggitrice*] of the house, upon whom the tranquility of the family depends," she informed him.

When Gregorio wrote her about an apartment that he had found close to the university, Bianca understood why the location might appeal to him, but it had the disadvantage, as she pointed out, of having fewer rooms than apartments located farther away. "If I was in Padua I could help you," she noted, but since that was apparently impossible, she was happy to leave the decision to him as the future "head of the household."[13] Months passed without signing a lease; by mid-July, with no apartment key yet in hand, Bianca suggested the couple should sing a *Te Deum* to mark their rejoicing when he had secured one. Her last letter to him as his fiancée, written on August 23, is full of news about wedding gifts, congratulatory nuptial sonnets (an Italian tradition), and visits from relatives and other well-wishers. A trunk of his clothing has arrived, and she is preparing her own traveling kit, filling it with needles, cotton balls, scissors, and buttons. She mentions a trip—most likely a honeymoon—but furnishes no details. And there's this good news: Ricci has finally found an apartment both of them are ready to live in as a couple.

They would move to larger apartments several times in the coming years as the family grew to include three children: Livia born in 1887, Cesare in 1891, and Giorgio in 1892. Many years later, Bianca's well-thumbed, annotated, and carefully preserved collection of homemaker's tips would be among the Ricci family papers donated to Lugo's public library, where it may still be seen today—a tribute to her role as family manager.

CHAPTER 6

# A Promotion That Wasn't

In Italy, as elsewhere, the arcane details of academic disputes were rarely considered front-page news fodder, but February 1887 in Padua proved to be an exception. On February 24, shortly after Ricci, professor *straordinario* (roughly equivalent to associate professor) of mathematical physics at Padua, forwarded his first three papers dealing with quadratic differential forms to the Lincei Academy's prize committee, an unsigned letter appeared on the front page of *La Venezia,* one of the region's widely circulated newspapers concerning a vacant position as professor *ordinario* (a full professor) in Ricci's department.

According to the anonymous author, it was an open secret among both locals and Padua's faculty members who sipped coffee and gossiped at Padua's cosmopolitan Caffè Pedrocchi, a resplendent two-story edifice across the street from the university, that a handful of Ricci's colleagues had quietly circumvented the rules during a meeting of the university's senior science faculty and robbed him of his just promotion to this professorship. Without naming the two promising young mathematicians vying for the position, the letter offered details of the meeting itself, including the offending "irregularity," and the subsequent complaint lodged with the university's rector by the professors who had stormed out of the meeting before the dean of the science faculty called for a vote. "The question is now pending at the ministry [of public instruction],"[1] the letter reported, which closed by encouraging its readers "in the meantime, to form their own opinions." Within a day or so, the newspaper had landed on Minister Michele Coppino's desk.

In his commemoration of Ricci that Tullio Levi-Civita prepared after his professor's death in 1925, he recalled that "promotion to full professor depended on a very small number of positions being available, and as a result there were great disparities"[2] among individuals, among whom he included Ricci. If Levi-Civita knew the convoluted backstory of his mentor's ill-fated efforts at promotion in 1887, this was the closest he ever came to acknowledging it openly. It was a saga that pitted the aloof, self-effacing Ricci, then thirty-four and embarking on work that would ultimately rank as one of the preeminent mathematical achievements of the late nineteenth century, against Giuseppe Veronese, a scrappy, politically savvy go-getter, more than a year Ricci's junior, and as Padua's professor *straordinario* of analytical geometry, no indifferent scholar himself. Before it was over, the question of this Padua promotion would reach all the way into the Italian parliament, many of whose members took a keen interest in what went on in Italy's national universities; and its final resolution would leave Ricci no closer to promotion than he had been at its start. But through it all, his methodical and painstaking work on the absolute differential calculus—now generally known, through its association with Einstein, as the tensor calculus—continued to advance.

Under Italian law, only applicants who had completed three years as professor *straordinario* were eligible for promotion into the most senior academic ranks, and Ricci dutifully submitted his first application to the ministry of public instruction at the end of 1883, almost three years to the day after his initial appointment. Veronese, a professor *straordinario* since 1881, turned his in, also with clockwork efficiency, in summer 1884, but the post for which both had applied went instead to a more senior mathematician, Francesco D'Arcais. Ricci seems to have accepted this decision without protest; as subsequent events would show, he respected the prerogatives of seniority. Veronese, however, objected fiercely, and appealed the decision to the ministry of public instruction, which oversaw such matters at the university level.

Veronese had a point. Unlike D'Arcais, whose slim publishing record the eminent Italian mathematician Francesco Tricomi once described as "not very much of particular importance,"[3] Veronese had already begun to establish his reputation with a paper he published in 1880, in German, in Klein's *Mathematische Annalen,* on projective geometry in higher dimensions. In it, he demonstrated how the subject could be organized and developed systematically from the point of view of geometry proper, rather than simply dressing up algebraic concepts in geometrical language. "It is clear that if this appointment can be held by the most senior [individual] or by others independent of merit, my promotion in the future will depend in large measure on chance and least of all on my scientific activity," he argued in his lengthy petition to Michele Coppino, the minister of public instruction. For good measure, he also included a list of his own publications (eleven, plus two at the printers, and one unpublished manuscript), a list of mathematicians, both Italian and foreign, who had cited his n-dimensional algebraic geometry in their own works, and his recent decision to participate in the 1883 competition for the Lincei's Royal Prize in Mathematics. Detouring into his financial situation, Veronese, who unlike the aristocratic Ricci hailed from a working class family and had been obliged to postpone his university education in his late teens to go to work, added, "I own nothing, I have no special teaching assignment, and my salary as *straordinario* must help my poor parents and my brothers."[4] That poignant personal detail may have resonated with Coppino, because several months later, after two years of stonewalling, the ministry finally made good on an earlier pledge to grant Veronese a new, permanent teaching assignment, apparently as a kind of consolation prize. He was appointed instructor for higher geometry at Padua, a course that brought him additional income and which he continued to teach for many years.

Thus, when the unexpected death in 1886 of Francesco Filippuzzi, who taught chemistry and the natural sciences at Padua, created a new opening for a full professor, Ricci and Veronese emerged as the obvious leading candidates. A third candidate, Giovanni Garbieri, professor *straordinario* of complementary algebra at the university, had begun his career as an elementary school teacher in Bologna, followed by a stint as a high school instructor and a succession of academic positions at technical institutes, before joining the science faculty at Padua.

Filippuzzi died in July, and the behind-the-scenes maneuvering for his vacated post began almost immediately. Five days later, Giovanni Canestrini, the dean of Padua's school of physical, mathematical and natural sciences, summoned the school's full professors to a meeting to nominate a successor. When they failed to agree on which candidate to rank first, all three contenders were ranked equally and in alphabetical order in the report that the dean submitted to Padua's rector, who, following the usual procedure in such situations, sent it on to Minister Coppino. In early August, two outspoken champions of Veronese in the Italian senate—Gustavo Bucchia, a one-time civil engineering professor at Padua, and Luigi Cremona, vice-president of the high council and director of Rome's Polytechnic School of Engineering—urged Coppino to approve Veronese's promotion himself. Bucchia, who wanted to see the candidates evaluated on their scientific merit, thought that on this basis Veronese deserved the nod. Cremona, known for his work on transformations of plane curves, the theory of birational transformations, and projective geometry, had longstanding ties to Veronese, who had finished his undergraduate degree in mathematics at Rome under his supervision and spent an additional four years there as assistant to the chair of projective and descriptive geometry. In a short but glowing note to the minister, Cremona insisted that no one—presumably he included Ricci in this category—could match "the powerful brilliance, creativity, and great originality" that Veronese possessed. He continued: "If any chair of geometry, pure or analytical, higher or not, were put to a *concorso* for a professorship, Veronese would win it one hundred percent, out of all the current professors *straordinari*. I nominate him then in the interest of science and of instruction, and I am convinced that the honorable Coppino, as he has so many other times, will also on this occasion accept my recommendation, which is not driven by any personal motive."[5]

Serving as minister of public instruction four times between 1867 and 1888, Coppino had earned his degree in letters and philosophy at the University of Turin and taught at several universities including Turin, where he served briefly as rector, before embracing a full-time political career in Rome in 1870. As minister, he initiated passage of an 1877 law that introduced free and compulsory schooling for children between the ages of six and nine years, pushed through the creation of a chair for Dante studies at Rome, and used the prestige of his office to carry through a national edition of Galileo's works.

Well-versed in Italian academic politics, Coppino replied to Cremona on August 13, promising to review the dossiers of all three candidates before the new academic year began. Ten days later, he sent a letter to Padua's senior science faculty, directing them to rank the candidates on merit, "*taking into account their respective scientific and teaching activities, and the importance of the chair held by each one* [emphasis in original]."[6] On November 9, five months after the jockeying for the Padua chair began, Dean Canestrini, a biologist largely remembered today for his support of Darwin's theory of evolution, convened an unprecedented, or as he termed it "exceptional," meeting of the school's full professors; once again, all ten attended. After formally presenting the minister's written charge to the gathering, the dean opened the floor for discussion. The first speaker, Domenico Turazza, director of Padua's engineering school, offered the opinion that Veronese's field was more important than Ricci's because of the large numbers of students taking Veronese's course in analytical geometry "as compared to the handful attending [Ricci's] lectures on mathematical physics"; later, in the same meeting, he disclaimed having any direct knowledge of Veronese's work. Another venerable member of the faculty, Enrico Nestore Legnazzi, added his voice to Turazza's comments about the greater importance of Veronese's field, not from any scientific considerations, he said, "but because it is fundamental for the other sciences."

A much younger professor, physicist Augusto Righi, expressed surprise that the faculty thought they could choose among the three candidates based on an examination of their published works when Minister Coppino himself relied on a task force of experts in the relevant scientific field to do the judging in comparable situations. Ernesto Padova, professor of rational mechanics and the author of many papers on elasticity and electromagnetism, also spoke up. A staunch friend of Ricci, he wondered how any of his colleagues could presume to set themselves up as a board of examiners. "Speaking for myself, I wouldn't know how to judge the relative value of these two colleagues, all the more so, as they work in two different branches of mathematics," he told them pointedly. Padova offered his own formula for judging the candidates:

> At first I wasn't going to speak about the importance of the chairs [i.e., Ricci and Veronese's respective fields], to avoid comparisons. But since others have spoken of it, I think that the criteria for the importance of a chair must be based on the luster that it gives to the university, on the sum of knowledge required of someone who is summoned to occupy it, and on the difficulty of finding the ideal individual to hold it with distinction. Now, by and large in the recent mathematical physics competitions, the

> applicants declared eligible were fewer than the available positions, while the opposite can be verified for the geometry competitions. Mathematical physics, then, requires more than any other science, a wealth of deep learning in related branches of mathematics. The reasons why the chair is important are then added to the importance of [Ricci's] seniority.

Did his colleagues, Padova went on, really believe that they had the technical acumen to judge the credentials of these particular candidates? In his opinion, they needed to form committees, perform an accurate examination of the relevant works, and provide a detailed, essentially unassailable justification for whomever they selected. "No one [here]," Padova asserted, "will feel able to assume this task."

Padova's remarks went unchallenged; and in the rush to find different, presumably more appropriate criteria, one faculty member asked how many votes each candidate had received when they had received their appointment as professor *straordinario.* Records indicated that Garbieri received thirty-nine votes out of fifty; Ricci, forty out of fifty; and Veronese, who had worked behind the scenes to make sure someone in his camp asked precisely this question,[7] forty-five out of fifty. When someone else pointed out that Veronese had recently earned top marks, fifty votes out of fifty, in competition for a chair of geometry at the University of Catania, in Sicily, Padova had a ready answer: "The votes obtained for the nomination as *straordinario* can't be considered, since they are reached by different commissions, probably informed by different criteria; besides, above all, we need to take into consideration for the promotion, what the *straordinari* have done since that original national competition." Furthermore since there had been no recent competition anywhere in Italy for a full professor of mathematical physics, there was no basis on which to make a meaningful comparison between Ricci and Veronese.

The meeting ended with two motions. The first—to decide the promotion on the basis of seniority (which would have guaranteed Ricci the job)—was voted on and tied, with five in favor, five opposed. The second motion, which carried by three votes, removed Garbieri from consideration, declared the faculty ill-equipped to rank the scientific credentials of the two remaining candidates, and unceremoniously deposited responsibility for the decision in Minister Coppino's lap, asking him to choose, "using criteria that he will believe to be more suitable."

The meeting may have been held behind closed doors, but word of its deliberations quickly reached both Ricci and Veronese. Soon afterward, Ricci contacted his own advocate in Italy's legislature, Ulisse Dini, now a member of the chamber of deputies (the lower house of parliament). Anxious to help, Dini arranged for his former student to meet Minister Coppino in Rome at the beginning of December. In that meeting, Coppino asked Ricci to provide him with a written account of "all the particular circumstances" surrounding his request for promotion. Once he had written it, Ricci asked Dini to pass it on to the minister, which he did. In an accompanying cover note, Dini recommended against involving Padua's faculty in the decision again and urged Coppino to push Ricci's appointment through, pointedly reminding him of Ricci's "greater seniority in relation to Professor Veronese," and adding, "I also believe that as matters stand now there is nothing else to do."[8] Dini's efforts on behalf of his protégé went nowhere; it was late December before Coppino responded, informing him that all matters pertaining to the promotion of Ricci, as well as Veronese, had ground to a halt.

There is no record that the minister replied directly to Ricci's petition. In it, Ricci had invited Coppino "to examine impartially" a list of issues raised by the November meeting, before pointing out that the minister's own criteria had succeeded in eliminating one candidate, but that the group had meanwhile declared themselves "incompetent to judge the relative scientific value" of the remaining two candidates, Ricci and Veronese. Turning his attention, once again, to the seniority issue, he wrote,

> Consequently, there does not exist... any reason for the faculty to suddenly treat this as a special case and depart from criteria consistently followed by it and by other departments and schools of application... in the proposals of promotion, especially dealing with professors *straordinari* elected by national competitions, that is to say, the criteria of seniority. If the undersigned has persisted and [continues to persist] on [adhering to] the pure and simple application of these criteria, it is not because he has reason to fear [being found wanting] on other grounds, but out of a desire always to avoid unpleasant comparisons that can only damage legitimate self-esteem and diminish one's moral standing in front of colleagues and students, as has been the case with others whose promotions have been postponed.

A proud Northern Italian, Ricci also expressed outrage at the implication that winning a competition for the position of full professor of projective and descriptive geometry at Catania (a university in Sicily!) entitled Veronese "to the same grade in the chair of analytic geometry occupied by him now in this university."[9] That Ricci knew what had occurred at the November meeting and was unhappily aware that Veronese's performance in the Catania competition had impressed many of the Padua faculty is clear from what he wrote next: "Nor can this same fact [i.e., the competition] constitute a basis for preferring Veronese over the undersigned, whether it is because this was [already] presented to the faculty in the course of the deliberations noted above, or because the undersigned had no opportunity to experience a *concorso* for the position of professor *ordinario* of mathematical physics."

Resting his case on these points, Ricci concluded his petition by making a "respectful request" for the minister "to arrange for the commission to be nominated, which will examine the new qualifications" for promotion.

What happened next both surprised and outraged Ricci and his supporters. As the front-page letter in *La Venezia* recounted, in mid-February 1887 Dean Canestrini had called a curious and unexpected meeting of his senior faculty. For a closer look at what transpired during these deliberations, we turn to a set of documents housed among the voluminous records of the ministry of public instruction in the State's Central Archives. The meeting, which had been announced simply as "Communications from the dean and possible deliberations following. Do not fail to attend, please," took place on February 19, during an official school recess. Only seven of the school's ten senior professors, four of whom had voted for Veronese at the November meeting, were able to attend. Once the group had assembled, Canestrini quickly called on Turazza to read a motion—which the aged engineer had prepared ahead of time—in favor of Veronese's promotion, which basically repeated all the arguments he had raised the previous summer. When they realized the trap set by Canestrini, the three members who had voted for Ricci in November demanded that the meeting be suspended; when the dean refused, Francesco D'Arcais (whose 1884 promotion Ricci, unlike Veronese, had refused to challenge), Giovanni Omboni, and Augusto Righi walked out of the room. The remaining four faculty members promptly voted to approve Turazza's motion.

When Ernesto Padova, who had been in Rome, learned what had happened, he wrote an indignant letter to the University of Padua rector, which D'Arcais, Omboni, and Righi also signed. Their letter, which they had requested be forwarded to Minister Coppino, languished in the rector's office for several days, perhaps more. On February 28, they took matters into their own hands and sent an angry letter directly to the ministry's headquarters in Rome. In it, they wrote: "This protest should have been transmitted to you [Michele Coppino], since all the papers relative to this affair, after having been held several days at the rector's by the wishes of the dean (who took himself to Rome immediately), were then sent to the minister, when the dean himself considered it appropriate."[10] In short, they argued that by creating "a fictitious majority" in favor of Veronese, Canestrini had manipulated the election.

Veronese's appointment had been approved in Padua on February 19; the letter about it to *La Venezia*, published on February 24, came to Coppino's attention the following day; and the meeting between Canestrini and Coppino in Rome probably took place soon after, almost surely before Padova's letter reached Coppino's office. These unexpected and presumably embarrassing public airings of confidential academic matters, and the ensuing upheaval among his own faculty, may have given Canestrini second thoughts about his none-too-subtle support for Veronese; in any case, at this February meeting, he offered his resignation to the minister.

An able and accomplished administrator, Coppino offered the dean one more chance to make things right with his faculty. That March, at a gathering of Padua's senior professors in the faculty of science, "conforming to the desire of the minister of public instruction,"[11] the penitent dean tried one final time to resolve the Ricci-Veronese promotion issue. Padova requested that the minutes of the controversial February meeting be read to the assembled group; Canestrini, who evidently felt that both his integrity and his authority remained under a cloud, obliged. He then proceeded to read aloud several other relevant documents, including his own letter of resignation to Coppino and Coppino's written refusal to accept it. Once again, Turazza called on the faculty to approve the promotion of Veronese; Padova responded with the by-now familiar motion to promote Ricci, reminding his colleagues that "a comparative judgment based on an examination of the works of the two candidates was never carried out either by this faculty or by competent persons elsewhere."[12] To no one's great surprise, the vote remained tied.

Later that spring, the politically savvy Coppino, perhaps to repay a favor or to gain credit for future favors, bowed to the senators who had tirelessly championed their candidate and promoted Giuseppe Veronese to professor *ordinario* of analytical geometry, a title he held until his death in 1917. Veronese himself was elected to the Italian senate in 1904, a position he also held for the rest of his life.

In his December petition to Coppino, Ricci had pleaded with the minister to resolve the promotion matter soon, in order to restore "the tranquility" he needed "to apply himself with all his might" to his [mathematical] studies and the teaching entrusted to him.[13] Now, in the wake of Veronese's promotion, Ricci looked to the minister for a sign that the words and praise offered on his behalf in the faculty's "long debate" would count in his favor in the future. As he wrote almost plaintively to Coppino, "I would be grateful if you would assure me that my teaching and my scientific work are also somehow approved by the minister's office and that in view of the votes of November 9, 1886 and March 3, 1887... you will keep me in mind for promotion as soon as it becomes possible."[14] Before drafting this letter, Ricci had consulted with Dini as to its contents. Dini, dismayed by the turn of events, had sought out Giovanni Ferrando, the head of the minister's cabinet, and the two of them had worked out how Ricci needed to couch his letter. Although Ricci's letter was addressed to Coppino, Dini entrusted it to Ferrando to hand it over to the minister, and attached the following note:[15]

> Here is the request... cast in the terms that we were speaking about some days ago. I would be grateful if you will respond to him [Ricci] saying that the minister, saddened by not being in a position to also promote him, and taking into account his merits, will initiate his promotion as soon

> as another position of full professor becomes vacant. Thus at least his self-respect will be preserved in part.

But Ricci had been deeply wounded. In September 1887, he joined five other candidates in applying for the vacant position of professor *straordinario* in infinitesimal calculus at the University of Modena, only to withdraw from consideration two months later. Another opportunity to leave Padua for an associate professorship—this one in higher mechanics at Bologna—arose the following year, and this time, Ricci thought of asking his mentor at the University of Pisa, Enrico Betti, for advice before doing anything. Hesitant to contact Betti directly, he wrote his friend and fellow mathematician Vito Volterra, himself recently promoted to full professor of rational mechanics at Pisa, asking him to speak to their former professor on his behalf. In his letter, Ricci explained why the offer was tempting: "Several on Bologna's faculty of science... realizing that a *concorso* [in higher mechanics] would probably not yield good results, would have voluntarily offered [the position] to me. For my part, I have always desired to be able to go to Bologna and all the more so after the wrong that has been done to me here."[16]

What did Betti think, he asked, about Ricci's taking a position that might not lead to a full professorship and where he might find himself in an inferior position with respect to other associate professors there? There is no indication of what Betti, or Volterra for that matter, thought. In any event, Ricci remained at Padua, where his long-sought promotion would remain beyond his grasp for the rest of the decade.

CHAPTER 7

# The Absolute Differential Calculus

> "I am well and always working on quadratic differential forms and similar things."
>
> G. Ricci to V. Volterra, Feb. 9, 1889

In the wake of his failed attempt to become a full professor, Ricci withdrew into his work, refining and developing the theory, methods, and notation for the subject he called the absolute differential calculus, today more generally known as tensor calculus. He also found solace in the birth of his daughter Livia—two sons would follow in 1891 and 1892—and in the classroom, presenting lectures aimed at generalizing the mathematical expressions of various physical phenomena. Ricci's lectures lacked theatrics, but they were compelling and famously precise, his student Levi-Civita later remembered. "Anyone who followed them attentively would have understood the core of the question being considered... and would have sensed the vigor of Ricci's lucid intellect."[1]

Putting that "lucid intellect" to the test, in 1887 Ricci submitted three papers on the theory of quadratic differential forms to a competition sponsored by the Lincei Academy for the Royal Prize in Mathematics—the most prestigious award open to Italian mathematicians of that era. Established in 1878 by King Umberto I,[2] the competition had attracted eleven entries, including one from Veronese, when it was first held in 1883 but was then put on hold for two years. Despite lengthy deliberations, or perhaps because of them, the commissioners ultimately decided not to make an award, but in their report, published in 1887 in the *Proceedings of the Lincei Academy* (*Rendiconti dell'Accademia dei Lincei*), they singled out the mathematical writings of Giulio Ascoli, Francesco Siacci, and Veronese for special commendation. They noted that the latter's work on $n$-dimensional algebraic geometry would have merited the prize if he had submitted other papers demonstrating how his research could be applied to a wider range of geometrical problems.

Meanwhile, the selection process for the 1887 prize had begun. The Lincei's commission—a coterie of Italy's most distinguished senior mathematicians, headed by Eugenio Beltrami—once again worked diligently, exchanging documents and other information by mail. Aside from Ricci, there were two other serious applicants:[3] Ascoli, competing for a second time, and Riccardo De Paolis, a colleague of Betti's at Pisa. The commission members were nothing if not thorough: It was April 1889 before Beltrami wrote to his fellow commissioner Enrico Betti, asking his long-time friend for his personal assessment of Ascoli's work. Beltrami trusted Betti's judgment. "If its positive value seems to you such that we would be able to consider it as deserving of a prize,"[4] he explained, "we will proceed to look more closely at it."

Apparently Betti thought otherwise, for in May, the president of the Lincei, Francesco Brioschi, met with Betti and another commissioner, Giuseppe Battaglini, in Rome and together they explored the possibility of dividing the Royal Prize between Ricci and De Paolis. Brioschi liked the idea, but Felice Casorati, Beltrami's colleague at Pavia, who served briefly on the commission, objected strenuously to the notion when Beltrami told him about it. Casorati, it seems, was worried that Betti might let slip something about the prize to De Paolis before a definite decision had been made. "For my part I am neutral,"[5] Beltrami admitted, before adding, "The reason for Casorati's opposition is the uncertainty in which he still finds himself about the intrinsic value of Ascoli's papers.... [He] is now reading the said papers and tells me that so far he has not found any truly important results; and he would really maintain that these papers can be classified below the other two."

By June 1889 Betti had received Battaglini's report and sent it on to Beltrami, who tried unsuccessfully to engage his own colleague in a conversation about the prize. As he wrote to Betti, "I would welcome talking about this with Casorati, but he has consistently refused, arguing that he sees no purpose to this commission."[6] Although Casorati was no longer on the panel, his negative comments may have helped persuade the remaining members to once again withhold the award. The scientific papers of De Paolis and Ricci "are only lacking some additional requirements to become worthy of the prize,"[7] Beltrami wrote, almost as an afterthought, in the official report he submitted to the Lincei after the commission's work had ended, perhaps intending to bolster the rejected aspirants' spirits and encourage them to persevere with their research.

Although the prize commissioners came up empty-handed for a second time, Ricci's consolation prize may have been Beltrami's authoritative commentary in that same report, on his three submitted papers. A formidable mathematician whom Ricci had first encountered as a student at Bologna, Beltrami was one of several geometers whose work Ricci had mastered by the time he began to publish the papers that in 1884 would usher in the intricate mathematical language today known as tensor calculus. In 1868 Beltrami's studies concerning the differential geometry of curves and surfaces had led him to the important discovery[8] that the abstract ideas and rules of Lobachevsky's geometry are a concrete realization of non-Euclidean geometry on surfaces of constant negative curvature. In an essay that traces Ricci's intellectual pedigree, the distinguished mathematician Dirk J. Struik writes that after extending his researches to three and more dimensions, Beltrami developed, following Riemann's thinking, a well-constructed theory of [constant curvature] spaces, "especially those of negative curvature, established a relation to non-euclidean geometry... and found the so-called *differential parameters,* a term he borrowed from [Gabriel] Lamé. He thus helped to create an apparatus for the systematic further exploration of differential geometry, in a way quite different from [Elwin] Christoffel and [Rudolf] Lipschitz."[9]

In the introduction to his 1884 paper "Principles of a theory of quadratic differential forms,"[10] published in *Annali di Matematica pura ed applicata,* Ricci explained why he considered this field of mathematics to be ripe for improvement:

> Almost all the geometers who up to now have been occupied with the course of ideas surrounding this subject, which has in fact been developed and introduced into the field of analysis quadratic differential forms representing the linear elements of space of $n$ dimensions, demand that the theory that applies to our space and for ordinary surfaces of two dimensions. be in accordance with their research. And if the added results are important, the methods don't always appear clear and don't give results as reasonable as those that come without questions to resolve.

Promising to use "purely analytical concepts" in his own research program, an approach that he believed offered the possibility of a more profound study of quadratic differential forms, Ricci proceeded to establish criteria for distinguishing between forms of class zero and one, while methodically demonstrating the necessary and sufficient conditions for a form to be of either class. In his report, Beltrami singled out Ricci for being the first to establish a rational classification of these forms; other geometers, he remarked, had already resolved this problem for forms of zero class, "but with the help of considerations not completely intrinsic to the subject: [Ricci] solves it completely by pure analysis in both cases—this one [forms of class 0] and the other one [forms of class 1] never previously studied and much less accessible."[11]

"The forms of the first class," the report continued, are important because "in the case of $n = 2$, they lead directly to the Gaussian starting point and for every other value of n, serve as the foundation for a generalized theory of surfaces in multi-dimensional spaces."[12]

Ricci's second entry, "On the parameters and invariants of quadratic differential forms,"[13] also published in the *Annali,* in 1886, offered a significant expansion of Beltrami's earlier pioneering work on the theory of differential parameters. In brief, Beltrami wrote, Ricci's 1886 paper "is devoted to the study of those expressions that in the theory of differential forms correspond to the invariants and covariants of algebraic forms."[14] Ricci instinctively approached the problem of differential invariants and differential parameters from an algebraic point of view; Beltrami, on the other hand, had invoked the calculus of variations in his approach to the study of differential forms. Their different mathematical styles may help to explain why Beltrami maintained a "neutral" stance in discussing the 1887 applicant pool with Betti.

More importantly, and with the benefit of hindsight regarding the role that absolute differential calculus was to play in Einstein's quest to express his theory of gravitation, Ricci's 1886 paper is significant because, as Italian science historian Fabio Toscano has pointed out, "the algorithm, already introduced by Christoffel—the covariant derivative—makes its first appearance"[15] in this paper "to generate the coefficients of a covariant quadratic differential form with respect to a given fundamental form." A simple example of covariance occurs in a simple harmonic oscillator in ordinary 2-dimensional space. The equations are

$$m\frac{d^2x}{dt^2} = -kx$$
$$m\frac{d^2y}{dt^2} = -ky,$$

where $m$ is the mass; $x$ and $y$ are the spatial coordinates; $t$ is the time; and $k$ is the spring constant.

If we rotate the plane by an angle $\theta$, it will have new coordinates related to the old ones by

$$x' = x\cos\theta + y\sin\theta$$
$$y' = -\,x\sin\theta + y\cos\theta.$$

When substituted into the equations above, this gives

$$m\frac{d^2x'}{dt^2} = -kx'$$
$$m\frac{d^2y'}{dt^2} = -ky'.$$

In other words, these are the same equations written into the new coordinates. These equations are then covariant with respect to rotations. The tensor calculus of Ricci and Levi-Civita offered a way of expressing covariance even for a quantity in four-dimensional spacetime.[16]

The possibilities inherent in this mathematical methodology prompted Ricci to prepare a brief description and explanation, aptly titled "On the covariant differentiation of a differential form,"[17] which Ulisse Dini presented on his behalf for publication in the Lincei's *Proceedings* in January 1887, shortly after the deadline for submitting material for the Royal Prize.

The third paper that Ricci submitted concerned systems of independent integrals of a linear and homogeneous first order partial differential equation, which drew on some of his own and Beltrami's earlier papers.

A consummate academic who knew how to deliver bad news to a younger colleague, Beltrami's 1887 Royal Prize commentary offered ringing praise of Ricci's "important studies" before reaching what Beltrami evidently considered the crux of the matter. Acknowledging that these studies "unquestionably constitute three works of great power and show the strong and varied analytical knowledge of their author,"[18] Beltrami then went on to ask

> whether the importance and the fruitfulness of [Ricci's] results correspond to the exertions he went through in order to reach them—a long, uninterrupted series of labored transformations that make the reader's task somehow painful and compel him not only to pay very close attention but also to have a profound knowledge of... higher analysis.

Writing on behalf of the entire commission, Beltrami concluded his remarks about Ricci's 1886 work with the following assessment:

> It seems to us to represent the preliminary development of a powerful tool that appears to have already led to interesting results. However, it still awaits its final justification by further tests in which perhaps its primitive and complex analytical apparatus could be replaced by simpler and more sophisticated approaches.[19]

In 1888, as deliberations over the Royal Mathematics Prize entered an inconclusive second year, Italy's oldest university, the University of Bologna, celebrated its eight hundredth anniversary, a date more or less plucked arbitrarily from the annals of history by the university's poet-in-residence, Giosuè Carducci. King Umberto came for the celebrations, as did dignitaries from around the world, along with a multitude of students who converged on the long narrow streets and dark arcades hugging the university quarter. Anticipating a feast, students from Turin brought a huge bottle of local wine strapped to a two-wheeled cart pulled by four oxen and conveying a student decked out as Bacchus. Their peers at the University of Pavia brought a gargantuan wheel of cheese decorated with lighthearted verses. Not to be outdone, Padua's students contributed a live ox, the emblem of the building (the *Palazzo del Bò*) that housed the university.

In a more sober vein, Padua's academicians marked the occasion with the publication of a series of papers by various faculty members. In his contribution, titled *Delle derivazioni covarianti e controvarianti e del loro uso nella analisi applicata* ("Covariant and contravariant differentiation and their use in applied analysis"), completed in March 1888, Ricci for the first time explicitly defined covariant and contravariant differentiation—operations he repeatedly encouraged other mathematicians to try—and then demonstrated how to substitute these two approaches for ordinary differentiation in problems ranging from the theory of surfaces to formulas relating to elasticity.

Two decades earlier, Christoffel, following in Riemann's steps, had investigated the conditions under which two quadratic differential forms can be transformed into each other. In the introduction to his classic paper *Über die Transformation der homogenen Differentialausdrücke zweiten Grades* ("On the transformation of homogenous differential equations of the second grade"),[20] published in 1869, Christoffel explained that "[his] researches originated in the $n$-dimensional extension of the problem of surfaces that can be developed onto one another."[21] He went on to construct two special differential expressions equipped with a system of super and sub scripts, a notation now known as Christoffel symbols. His results (which Felix Klein later named the Christoffel reduction theorem) and Riemann's concept of space "constitute the premises of the absolute differential calculus," Levi-Civita later wrote, adding, "Its development as a systematic branch of mathematics was a later process, the credit for which is due to Ricci, who during the years 1887–1896 elaborated the theory and worked out the elegant and comprehensive notation that enables it to be easily adapted to a wide variety of questions in analysis, geometry, and physics."[22]

According to Struik—who learned his tensor mathematics from the Dutch mathematician Jan A. Schouten in Delft in 1917 and met Ricci several years later in Padua—Ricci took Christoffel's paper on the general problem of the transformation of one differential equation into another as his own starting point. "It is Christoffel's [1869] paper that served as the direct transmission belt between Riemann and Ricci," Struik later remarked in a volume honoring Riemann's legacy.[23] He added, "Christoffel's aim [was] to find the conditions of equivalence of differential forms, which happened to lead him independently of Riemann to the mathematical ideas underlying his lecture of 1851 [on the foundations of geometry] and to lead in a direct way to the Ricci calculus."[24] Ricci's earliest writings on this topic, he concluded, "show [him to be] a student of Christoffel's paper, even using the Christoffel symbols in Christoffel's notation."

Ricci's 1888 paper on covariant and contravariant differentiation and their use in applied analysis marked a turning point in the development of the concepts and notation of his calculus. The paper opened by reviewing the technique Christoffel had used in solving the equivalence problem of two quadratic differential forms. Developing Christoffel's ideas, he noted, had led him to modify the customary procedures used in the usual differential calculus. The procedure itself he called covariant differentiation. In Ricci's hands, the formulas and results always remained independent of the choice of variables, making it, he asserted, a valuable tool for applied mathematicians.

The question of what to call his calculus gave Ricci pause. In his 1888 paper, he described his invention as "systems of functions" or simply "systems." From then on, according to the Italian historian of science Luca Dell'Aglio, who has studied the origins of the concept of covariant differentiation, "in every systematic treatment of his methods, Ricci would always designate a differentiation defined on a particular system of functions as covariant differentiation."[25]

In contrast to his hesitation over what to name his new methodology, Ricci was quick to enumerate its benefits. In his words,

> The differential equations we obtain through the application of analysis to geometry, to mechanics, and to physics are necessarily independent of the choice of variables. Such a degree of independence cannot easily be inferred from the notation commonly used, because it does not take into account the form that the selected variables give to the expression for the line element of space. The possibility of taking into account this form depends, as we will see, on the fact that the equations with which we are dealing precisely because of their independence of the choice of variables, always contain systems of functions that in their totality can be represented by symbols... in which each index is open to all values from 1 up to 3.[26]

He then explained how the equations usually employed to describe the components of a force, a torque, heat conduction, or the displacement of a point could be recast in the form of "systems of functions."

In the next section of the paper, Ricci described the reciprocal operations of covariant and contravariant differentiation, the rules to follow in using his tools. Just as an artfully designed sculpture exhibit enables viewers to study the figures on display from every possible angle, Ricci's calculus allowed a geometric or physical system to be analyzed from many points of view, represented by a system of coordinates. The systems appear in the equations in ways that are independent of particular coordinate systems, but the values of their components change, depending on coordinate choices. "The equations," Ricci further explained,

> can or could be laid down directly in general coordinates with the application of one of the two operations [of covariant and contravariant differentiation], or after having established it in orthogonal cartesian coordinates... be transformed into any coordinates... In my opinion, in fact, these methods lead to a more legitimate extension of research relating Euclidean space in three dimensions to manifolds in $n$-dimensions, and of any nature. This is true because, considering that such extensions lack any experimental foundation * what mostly emphasizes their meaning is the purely analytical analogy between covariant or contravariant systems obtained by covariant differentiation with respect to different quadratic differential forms providing the results of the geometric, mechanical,

* Until Einstein, of course.

> or physical interpretations that [these methods] can suggest remain a free and independent process.[27]

Ricci devoted the rest of his paper to recasting formulas in differential geometry, elasticity, and heat conduction into systems of covariant and contravariant forms, convinced that if shown the way, others would then see the advantages that he saw in his new calculus.

His first readers were not so easily persuaded. Although published too late to be entered in the mathematics prize competition, Ricci's paper swiftly drew the attention of Beltrami, who in his 1889 commentary assessed the "technical feasibility"[28] of using the author's tools in handling problems in geometry and mathematical physics. These methods, Beltrami noted, depended on "a certain system of analytical procedures, called by him [Ricci] covariant and contravariant differentiation." Acknowledging that Ricci had already utilized these techniques in the three papers submitted for the Royal Prize competition, Beltrami maintained that he had nevertheless neglected to spell out clearly and precisely the machinery underlying those special methods. As for the benefits that Ricci confidently predicted would flow from these new methods, it was Beltrami's view that these applications had yet to "reach a point that would have put in clearer light the utility of the new algorithms."[29]

In the biographical memoir he prepared after Ricci's death in 1925, Levi-Civita diplomatically described Beltrami's "instructive" commentary as "an appreciative but cautious assessment" of Ricci's work, citing also "[Beltrami's] concerns about whether this new tool would lead to deep new results in the future."[30] Shortly after his public reading of the memoir, Levi-Civita received a letter from his colleague Francesco Porro, an astronomer in Genoa, who complimented Levi-Civita on his skill in tactfully treating Beltrami's comments about Ricci's studies. He said it reminded him of a remark he had personally heard Beltrami make in 1882 or 1883 to the effect that James Clerk Maxwell's "powerful *intuitive* mind" had ignited a revolution of "exceptional value" thanks to his electromagnetic theory; but that [Beltrami] refused to marshal the [mental] energy necessary to master the mathematical methods used by the great Scotsman."[31] Porro continued: "If we line up Ricci's algorithms with those analogous to the methods founded on quaternions [a complex set of numbers invented by W. R. Hamilton to characterize a four-dimensional vector space], it is perhaps not an exaggeration to say that Beltrami's great mind was affected by a curious idiosyncrasy that I don't feel myself sufficiently mathematical to examine." Porro's letter ended with a classic Latin injunction against passing judgment in areas outside one's expertise: *Ne sutor ultra crepidam* (literally, "shoemaker, don't gaze beyond the sandal").

It was not until 1892 that the first Royal Mathematics Prize was finally conferred on Luigi Bianchi and Salvatore Pincherle, two well-established and bona fide members of the Lincei Academy. The commissioners seemed genuinely disappointed that Veronese, who had entered the competition for a second time and included in his submission a manuscript for a voluminous textbook on the principles of higher geometry, had not followed the advice given him in the earlier competition. As the report noted, "We are not able to confer the Royal Prize for a work that does not contain a substantial discovery of new and important scientific truth."[32]

A year later, in 1893, Ricci finally settled on a name for the mathematical apparatus he had painstakingly constructed over the course of a decade. In a French-language account of his methods, published in 1892, he referred to it as the "calcul différentiel invariantif," the differential invariant calculus. In a paper published the following year, he adopted another name, which he described in a footnote that read: "For the sake of brevity I will denote by the name 'absolute differential calculus' (*Calcolo differenziale assoluto*) the totality of methods that I named on other occasions covariant and contravariant differentiation, because they can be applied to any fundamental form independently of the choice of the independent variables and require instead that the latter are completely general and arbitrary."[33] This was the same methodology that was to rivet Einstein's attention early in the next century.

Years later, Tullio Levi-Civita would retrace, in largely non-technical language, the intricate mathematical road that Ricci followed on his way to the absolute calculus. He began by recalling how his professor "gradually seized upon and perfected the calculations" in his first writings on differential forms, differential parameters, and partial differential equations and then "modified the usual procedures employed in the differential calculus in such a way that the formulas and results always remained in the same form, no matter what system of coordinates was being used."[34] He continued, "This explains why we have... a system of functions [now called tensors] that behave in the same way when the coordinates are changed, independent of the choice of these coordinates. In addition, certain operations are introduced that are equally independent of the coordinates chosen, i.e. they are absolute, giving the name to the calculus." Turning next to the advantages offered by Ricci's methods, Levi-Civita listed some of "the most useful applications [that] arise when the nature of the material under consideration requires a quadratic differential form," as in the case of general relativity, where the very small "interval between two events in space-time" is expressed using a differential form. At that point, explained Levi-Civita, Ricci took "this differential form as given—that is, as absolute—and this is where the essential element of the new calculus arises: in the notion of covariant differentiation, which has the essential characteristics of ordinary differentiation, but also respects the invariant behavior (i.e., is independent of the choice of coordinates) of the system to which it is applied."

Ricci's absolute calculus attracted little notice in the Italian mathematical community at the time of the 1892 Royal Prize award, nor did Ricci enter the next competition in 1895.

At a remove of more than a century, and biased as we are by the knowledge of its tremendous utility in Einstein's hands, it is hard to reconstruct precisely what Ricci saw in his work that for the most part escaped his contemporaries. Perhaps he too was being utilitarian in emphasizing the value of coordinate invariance for its simplicity and versatility, whereas Einstein was to give it a profound physical meaning. But at the time Ricci, then anxiously anticipating a new round of efforts to finally secure his university promotion, had no inkling of that. Is it too fanciful to suppose that as the nineteenth century drew to a close, even a personality as aloof and generally disdainful of academic intrigue as his looked at the body of his work and at the recognition (and perhaps even the easy cordiality) enjoyed by many of his mentors and peers and wondered if such regard would ever come his way?

**Photo 1.** Ricci Curbastro family tree
*Image courtesy of Archivio delle Ancelle Sacro Cuore di Lugo*

**Photo 2.** Gregorio Ricci Curbastro (ca. 1869)
*Photo courtesy of Ida Ricci Curbastro/Biblioteca Comunale Fabrizio Trisi, Lugo*

**Photo 3.** Postcard scene, Lugo (ca. 1900)
*Photo courtesy of Biblioteca Comunale Fabrizio Trisi, Lugo*

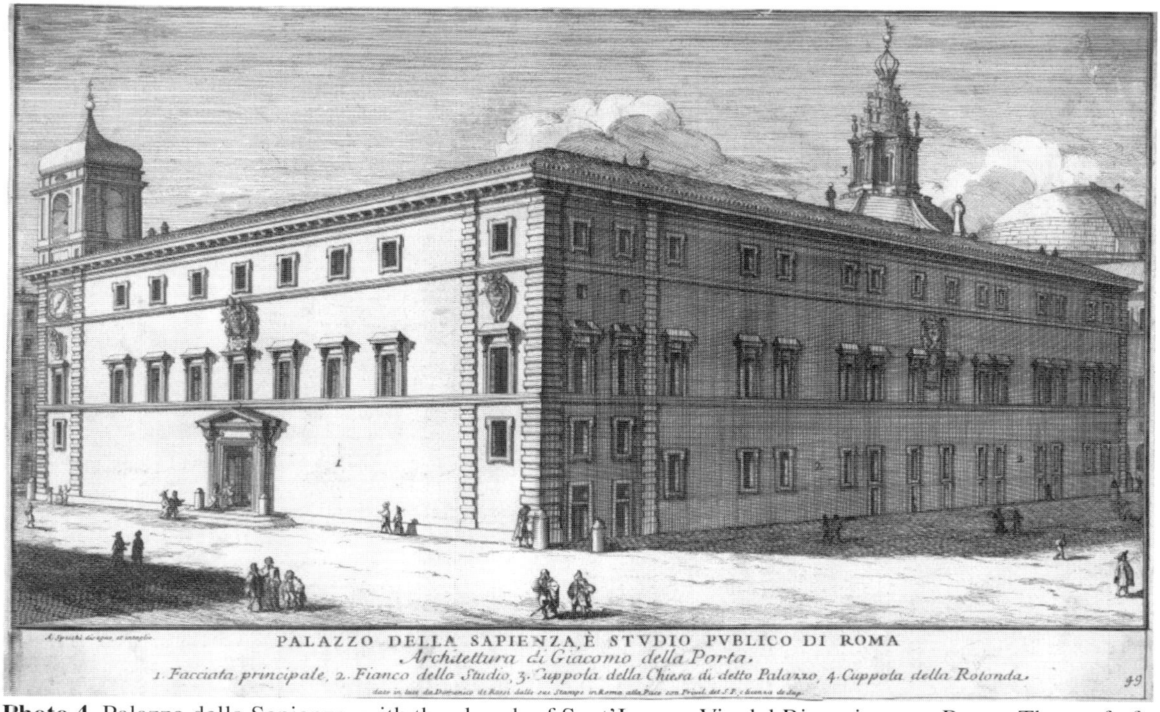


**Photo 4.** Palazzo della Sapienza, with the church of Sant'Ivo, on Via del Rinascimento, Rome. The roof of the Pantheon is visible in the background.
*Image is from a private collection*

**Photo 5.** Enrico Betti (1823–1892)
*Photo courtesy of Scuola Normale Superiore Pisa*

**Photo 6.** Ulisse Dini (1845–1918)
*Photo is courtesy of Institut Mittag-Leffler*

**Photo 7.** Eugenio Beltrami (1835–1900)
*Image courtesy of Archivio dell' Accademia Nazionale dei Lincei, Fondo Vito Volterra*

**Photo 9.** Ernesto Padova (1845–1896)
*Photo courtesy of Archivio dell'Accademia Nazionale dei Lincei, Fondo Vito Volterra*

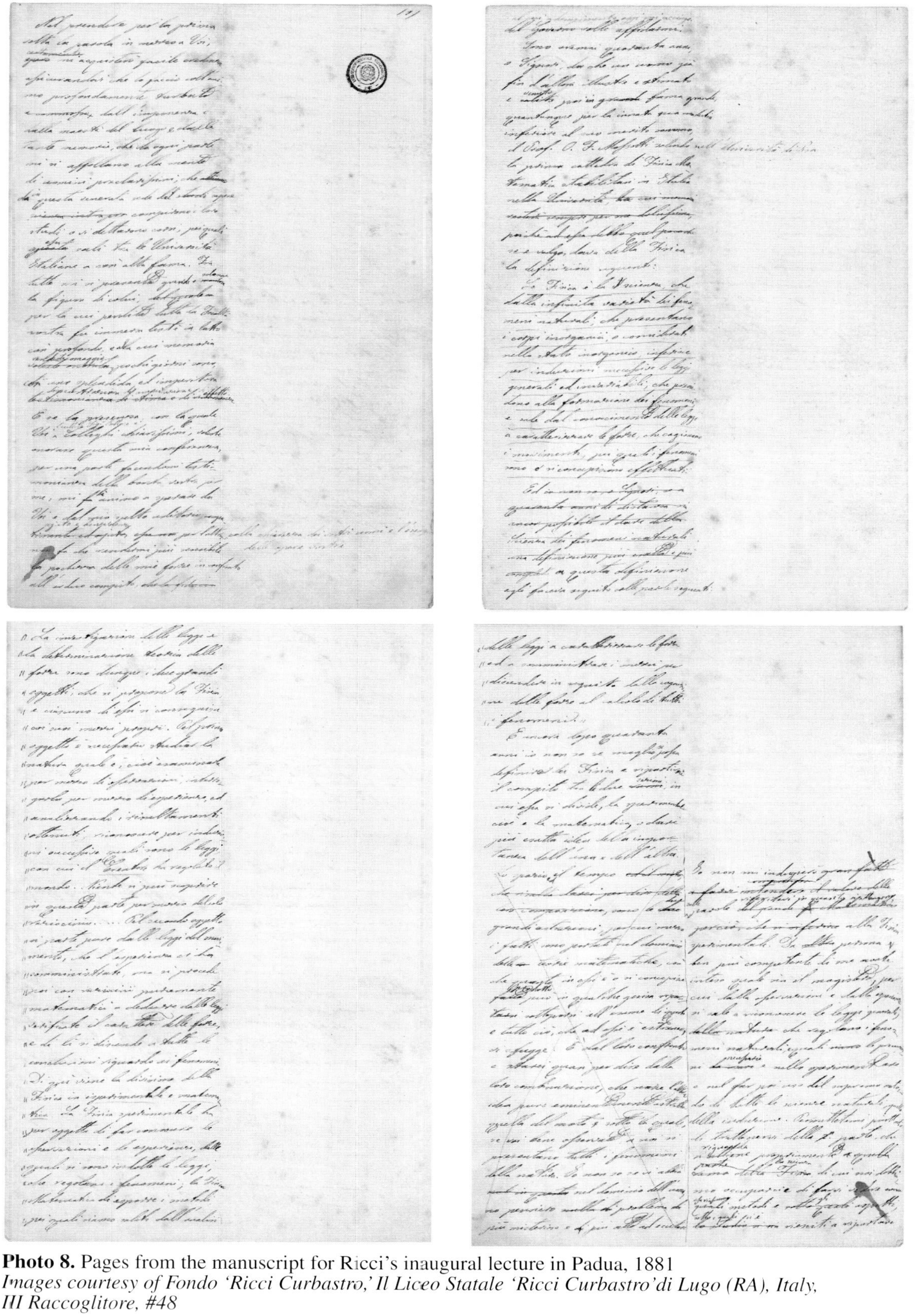

**Photo 8.** Pages from the manuscript for Ricci's inaugural lecture in Padua, 1881
*Images courtesy of Fondo 'Ricci Curbastro,' Il Liceo Statale 'Ricci Curbastro' di Lugo (RA), Italy, III Raccoglitore, #48*

**Photo 10.** Bianca Bianchi Ricci Curbastro
*Painting courtesy of Maria Giovanna Ricci Curbastro/Biblioteca Comunale Fabrizio Trisi, Lugo.*

**Photo 11.** Bianca Bianchi Ricci Curbastro
*Photo courtesy of Maria Giovanna Ricci Curbastro*

**Photo 12.** Gregorio Ricci Curbastro (1852–1925)
*Photo courtesy of Archivio dell'Accademia Nazionale dei Lincei, Fondo Vito Volterra*

**Photo 13.** Giacomo Levi-Civita (1846–1922)
*Photo courtesy of ASSR, Senato del Regno, Segreteria, fascicolo personale del senatore Giacomo Levi-Civita, n. 1274.*

**Photo 14.** Signed portrait of Garibaldi, with the following dedication, "To Giacomo Levi, G. Garibaldi [Giacomo had not yet added the maternal surname]"
*Photo courtesy of Ceccherini-Silberstein Family*

**Photo 15.** Giuseppe Veronese (1854–1917)
*Photo courtesy of Wikimedia Commons*

**Photo 16.** Luigi Bianchi (1856–1928)
*Photo courtesy of Wikimedia Commons*

**Photo 17.** Felix Klein (1849–1925)
*Portrait courtesy of Archivio dell'Accademia Nazionale dei Lincei, Fondo Vito Volterra*

# Méthodes de calcul différentiel absolu et leurs applications.

Par

M. M. G. Ricci et T. Levi-Civita à Padoue.

## Table des matières.

### Chapitre I.

**Algorithme du calcul différentiel absolu.**



### Chapitre II.

**La géométrie intrinsèque comme instrument de calcul.**



### Chapitre III.

**Applications analytiques.**



**Photo 18.** Ricci-Levi-Civita's paper was published in the first issue of volume 54 of the *Mathematische Annalen*, issued in October 1900.
*Image courtesy of Cushing/Whitney Medical Library, Yale University.*

**Photo 19.** Hermann Minkowski (1864–1909)
*Photo courtesy of Archivio dell'Accademia Nazionale dei Lincei, Fondo Vito Volterra*

**Photo 20.** Marcel Grossmann (1878–1936)
*Photo courtesy of ETH-Bibliothek Zürich, Bildarchiv / Fotograf: Unbekannt / Portr_01239 / Public Domain*

**Photo 21.** Albert Einstein (1879–1955)
*Photo courtesy of ETH-Bibliothek Zürich, Bildarchiv/Fotograf: Unbekannt/ Portr_03101/Public Domain*

**Photo 23.** Tullio Levi-Civita (1873–1941) (ca. 1912)
*Photo courtesy of Ceccherini-Silberstein Family*

**Photo 24.** Max Abraham (1875–1922)
*Photo courtesy of the Archives, California Institute of Technology*

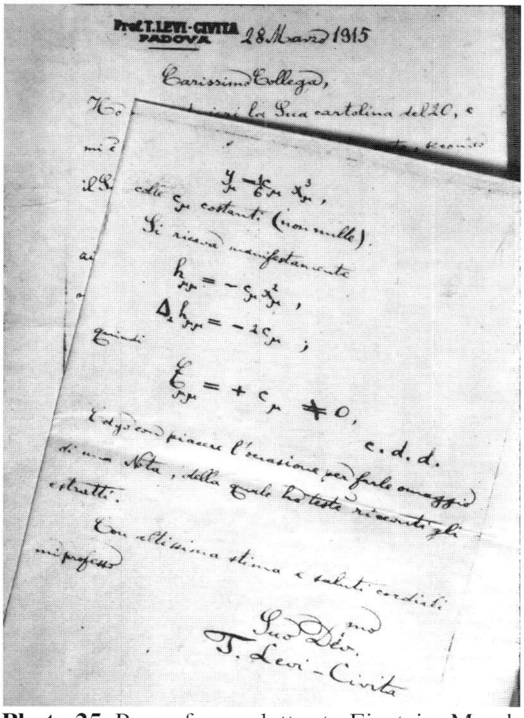


**Photo 25.** Pages from a letter to Einstein, March 28, 1915
*Photograph by Carl Glassman © The Hebrew University of Jerusalem, Israel*

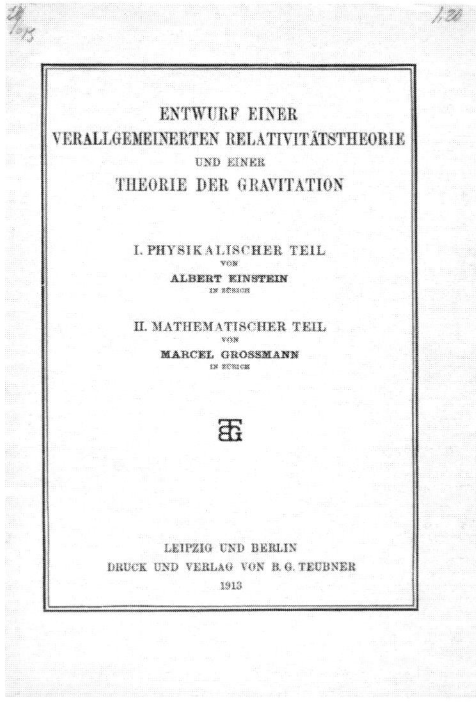

ENTWURF EINER
VERALLGEMEINERTEN RELATIVITÄTSTHEORIE
UND EINER
THEORIE DER GRAVITATION

I. PHYSIKALISCHER TEIL
VON
ALBERT EINSTEIN
IN ZÜRICH

II. MATHEMATISCHER TEIL
VON
MARCEL GROSSMANN
IN ZÜRICH

LEIPZIG UND BERLIN
DRUCK UND VERLAG VON B. G. TEUBNER
1913

**Photo 22.** Title page of the "Outline [or preliminary version] of a Generalized Theory of Relativity and of a theory of Gravitation," published by Einstein and Grossmann in early summer 1913. The paper contains Einstein's first application of Ricci and Levi-Civita's tensor calculus.
*Image courtesy of Paul Epstein Papers, the Archives, California Institute of Technology.*

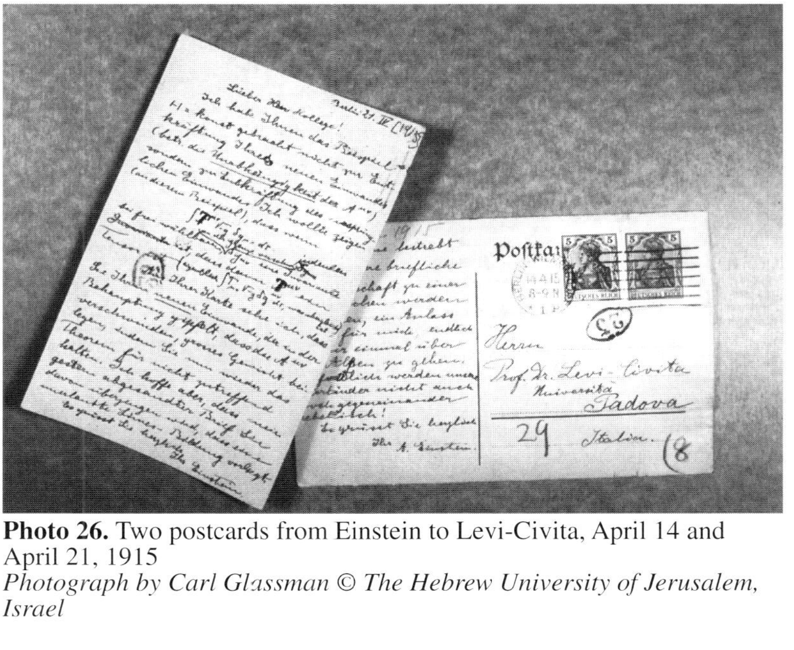


**Photo 26.** Two postcards from Einstein to Levi-Civita, April 14 and April 21, 1915
*Photograph by Carl Glassman © The Hebrew University of Jerusalem, Israel*

**Photo 27.** Caffe Pedrocchi, Padua (ca. 1915)
*Postcard scene from a private collection*

**Photo 28.** Passport for Libera Levi-Civita, issued in Padua, Sept. 30, 1918
*Photo courtesy of Ceccherini-Silberstein Family*

**Photo 29.** Gregorio Ricci Curbastro (1852–1925)
*Photo courtesy of Ceccherini-Silberstein Family*

**Photo 30.** Gregorio Ricci Curbastro and family,
Sant'Agata sul Santerno (ca. 1921)
Front Row: Gregorio Ricci (first on the left), Bianca (Livia's daughter, third from left), Giorgio (Gregorio's son). Second row: Remigio Lanzoni (Livia's husband), Livia (Gregorio's daughter), Giulia Archi (Cesare's wife,), Cesare (Gregorio's son).
*Photo courtesy of Archivio Fotografico, Biblioteca Comunale Ricci Garotti, Sant'Agata sul Santerno, RA, Italia*

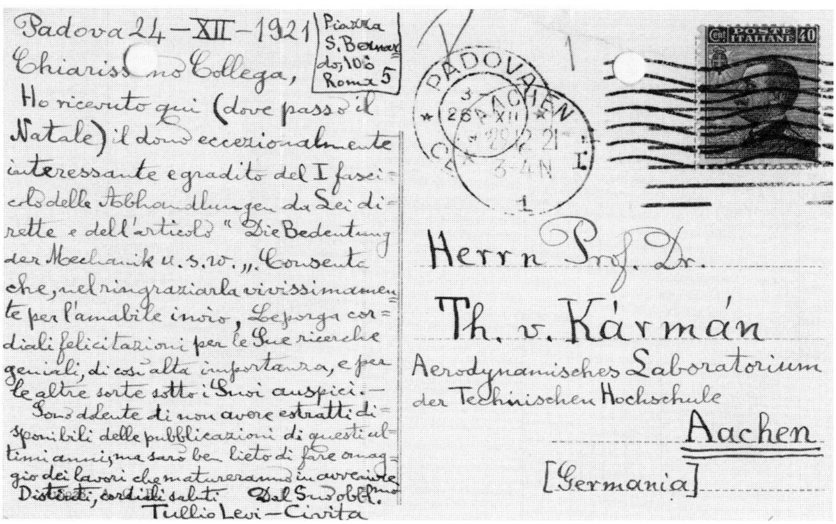

Padova 24–XII–1921 Piazza S. Bernardo, 10 Roma 5
Chiarissimo Collega,
Ho ricevuto qui (dove passo il Natale) il dono eccezionalmente interessante e gradito del I fascicolo delle Abhandlungen da Lei dirette e dell'articolo "Die Bedeutung der Mechanik u.s.w.". Consenta che, nel ringraziarla vivissimamente per l'amabile invio, Le porga cordiali felicitazioni per le Sue ricerche geniali, di così alta importanza, e per le altre sorte sotto i Suoi auspici.— Sono dolente di non avere estratti disponibili delle pubblicazioni di questi ultimi anni, ma sarò ben lieto di fare omaggio dei lavori che matureranno in avvenire. Distinti, cordiali saluti. Dal Suo dobbl.mo
Tullio Levi-Civita

Herrn Prof. Dr.
Th. v. Kármán
Aerodynamisches Laboratorium
der Technischen Hochschule
Aachen
[Germania]

**Photo 31.** Postcard from Levi-Civita to Theodore von Kármán
*Photo courtesy of the Archives, California Institute of Technology*

In 1921, Levi-Civita, while thanking von Kármán for a reprint, also praised him for his involvement in the European Applied Mathematics and Mechanics Movement. Von Kármán then organized the 1922 conference on hydro- and aerodynamics in Innsbruck, Austria, which although boycotted by the French and British, attracted 33 applied mathematicians and physicists from seven countries. Von Kármán and Levi-Civita divided the organizing costs.

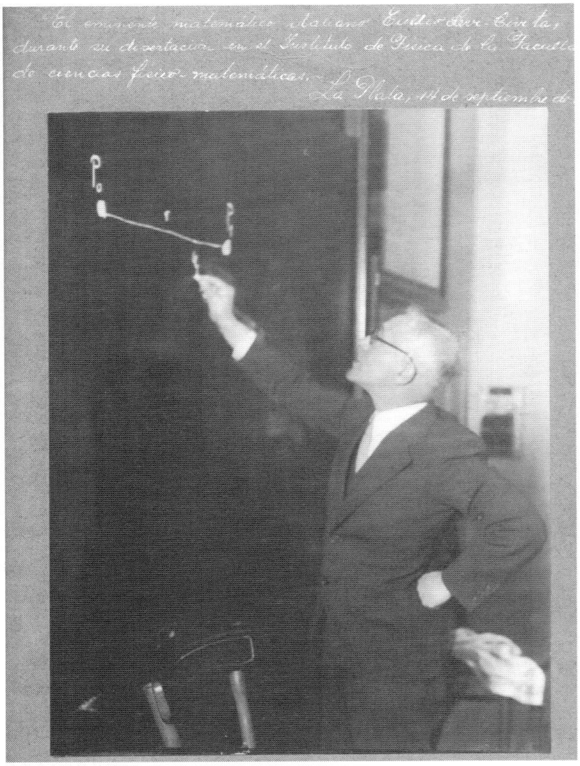


**Photo 32.** Tullio Levi-Civita, La Plata, Argentina, 1923
*Photo courtesy of Ceccherini-Silberstein Family*

**Photo 34.** Tullio Levi-Civita and his wife Libera Trevisani prepare to disembark in New York City, 1933
*Photo from a private collection*

**Photo 33.** A humorous depiction of Levi-Civita revealing the mysteries of the absolute differential calculus (ca. 1923). The student in the second row is thought to be Enrico Persico, who edited the Italian edition of Levi-Civita's calculus book in 1925.
*Image courtesy of Enrico Persico Papers, Dept. of Physics, Sapienza University, Rome*

CHAPTER 8

# The Alter Ego

In the summer of 1890, one hundred and three high-school seniors sat for final exams at the Liceo Ginnasio Tito Livio, one of the oldest and most prestigious public schools in the city of Padua. The subjects included oral and written exercises in Italian literature and letters, as well as translations of passages from Latin and Greek to Italian, followed by more proficiency tests in history, geography, natural history, philosophy, physics, and chemistry. Topics in geometry and algebra rounded out the suite of subjects. The tests were both exhaustive and grueling; seventy-three students are recorded as having failed at least one of them *per aver interrotto le prove,* meaning they didn't make it to the end of the exam. Given the option of choosing a written exam in Greek or mathematics, only three students chose mathematics. Among them was the thirteenth candidate on the alphabetical roster, at seventeen one of the youngest in the group, and he passed all his tests—a total of eleven oral and written exams—earning a near perfect score of 107 out of a possible 110, a record few, if any, students came close to matching in that year or any other. Two generations removed from the confines of the ghetto, where nearly all Italian Jews had been sequestered for centuries, this formidable young intellect, Tullio Levi-Civita, was destined to leave his mark not only on numerous branches of mathematics but also to influence the future course of science in a manner no one could have foreseen at the end of the nineteenth century. A mathematician of enormous versatility, he almost certainly remains best known for his work on the absolute calculus and the prominent role it played in Einstein's formulation of the general theory of relativity.

Tullio Alessandro Levi-Civita was born on March 29, 1873, in Padua, and grew up in the shadow of its university, where he would teach for twenty years. Guidebooks of that era describe a picturesque city of about 66,000 inhabitants, including an ancient Jewish quarter, and many bridges, some dating from Roman times, over various branches of the Bacchiglione River flowing through the town. What the city may have lacked in modern industrial establishments at that time—contemporary records allude to a single clothing factory (a vestige of a once-thriving wool industry) with seventy workers and a lone foundry with a hundred employees—it made up for in the intensive cultivation of agricultural products and its geographical position as a hub for the growing national railway and highway system that connected southern and central Italy with the northeastern part of the country.

No thread running through the fabric of life in the city, the historian Angelo Venture has noted, was more important than the university, which served as "the cultural metropolis"[1] of the Veneto, the region bordering the Dolomite mountains and the Adriatic Sea. Padua's "impetus to innovate [and] the push for industrial and financial initiatives" arose from the fusion of intellectual talent within the university's walls—"above all the technical-scientific knowledge of its mathematical faculty and the school of applications for engineers"—with the city's Jewish middle-class professionals, for whom emancipation in 1860 had unleashed a flood of new civic, commercial, and real estate opportunities.

The rolls of the Università Israelitica di Rovigo record the birth of Tullio's father, Giacomo Levi, the son of Abramino Levi and Rachele Civita, in 1846, and the further notice that in December 1868, Giacomo added his mother's surname to his own. Family members recall that Tullio's father, a lawyer, jurist, and politician, wanted to distinguish himself from another Giacomo Levi, no relation, who also practiced law in the same city; and perhaps he also wished to avoid confusion with his brother-in-law, yet another Giacomo Levi, who was director of the insurance company *Assicurazioni Generali,* in Venice.

Padua and its surrounding territories were part of the Austrian empire at the time of Giacomo's birth. Family tradition holds that he had, by the age of thirteen, become so vocal in his opposition to Austrian rule that his parents withdrew their young firebrand from the local Regio Ginnasio liceale S. Stefano[2] (the name of the school was changed to Tito Livio in 1872), where he ranked first in his junior high school class. He was sent to study 250 miles to the west in Piedmont, the birthplace of the movement for Italian independence, where, his family may have felt, an outspoken young Italian Jew could agitate in relative security. There he earned his laurea in jurisprudence at the University of Pavia and served as a volunteer with the charismatic general Giuseppe Garibaldi in two of the military campaigns that led to the unification of the Italian peninsula under Italian rule. He subsequently became mayor of Padua and a senator of the Kingdom of Italy.[3]

His son, Tullio, inherited his father's marked liberal bent, and although in his youth he played no active role in politics, he had the highest regard for his father and the institutions and liberties he stood for. In later years, two portraits would hang in his study at Via Sardegna 50 in Rome: one of his father and the other of Garibaldi, bearing the following dedication. "To Giacomo Levi, G. Garibaldi [Giacomo had not yet added the maternal surname]." It has been said by his mathematics colleague Ugo Amaldi that Levi-Civita inherited his strength of character from Giacomo and his capacity for empathy from Bice Lattes, his mother, whom he adored.

The house where Levi-Civita was born still stands on Via Daniele Manin 7, a narrow thoroughfare that runs between the bustling market square and the Piazza del Duomo, bordering the former Jewish ghetto. The city's Jews had been confined to their own quarter, a warren of dark, narrow streets and unhealthy homes, for nearly two centuries, until the doors that locked them in every night were torn down in 1797 following the arrival of Napoleon's troops in the city. During Levi-Civita's childhood, the city had a small Jewish population of around 950, which declined steadily in the following decades. The Italian journalists Annie Sacerdoti and Luca Fiorentino report that 600 Paduan Jews were still living there in 1938 when the Fascist government introduced the infamous racial laws. Of these, they report in a melancholy coda, forty-seven were subsequently deported, and "only three returned at the end of the war."[4] In matters of religion, Levi-Civita later described himself as an agnostic; in a detailed curriculum vitae that he filled out for university officials in the 1930s, he made a column, labeled it "race and religion," and underneath wrote, "Jewish Non-practicing [*Ebraica Aconfessionale*]."[5]

This strain of secular thought seems to have run through his immediate family. Until the age of ten, Levi-Civita was tutored privately by Luigi Padrin, an erudite priest with an encyclopedic knowledge of medieval Padua's literary history. Like his father before him, at age ten he was enrolled by his parents in the highly regarded Tito Livio, one of the oldest and most important schools in the city. Having apparently skipped a couple of grades, he entered the second year of junior high in 1883 and plunged into the demanding academic curriculum set down by the ministry of education. In 1886–1887, during his final year as a middle-school student, Levi-Civita's weekly schedule consisted of seventeen hours of Greek, Latin, and Italian; three hours of history and geography; two hours devoted to elements of natural history, and one hour of practical arithmetic. Also housed in the registrar's files is a record of his conduct and grades, with the note "*Licenziato*," meaning that he obtained the *licenza* or diploma to enroll in the high school in the fall, with the added distinction of graduating first in his class.[6]

Levi-Civita embraced even more subjects in high school, ranging from mathematics and philosophy to nine hours a week of chemistry and physics, which he took during his senior year. A slight figure with an ebullient personality, the budding mathematician was popular with his boisterous classmates, both on the sports field and in the classroom, where his ability to effortlessly understand and assimilate new material fascinated the other pupils. Sixty years later, a schoolmate by the name of Felix Carli, who had in the meantime become a seismological engineer in Argentina, remembered the glee with which a group of Levi-Civita's school friends delighted in chanting verses composed by "my dearest Tullio,"[7] of which he could still recite the following fragment: "And when we are done with the chemistry exams... because then we really are through!" (Unfortunately his memory could no longer supply what were evidently the crucial middle lines of the verse.) In 1888, the number theorist Paolo Gazzaniga joined the high school's faculty. By then, Levi-Civita's maternal uncle, a retired engineer, had introduced his nephew to the classic geometry texts of the English mathematicians Isaac Todhunter and George Salmon.

Eager to put his new-found knowledge to work, Levi-Civita set out to prove Euclid's puzzling fifth postulate, the parallel postulate, using only the first four postulates. In later years, Gazzaniga never grew tired of recalling how his extraordinarily precocious student confidently proceeded in a flawless and elegant way—only to arrive at a proof containing a circular argument. "At fifteen, in the disappointment of that failure, Levi-Civita could certainly not have imagined that one day his name would be attached forever to an extension, both inspired and potent, of that same concept of parallelism,"[8] Ugo Amaldi, with whom Levi-Civita would one day write a series of textbooks, remarked in his commemoration of Levi-Civita's life in 1946. Teacher and pupil would cross paths again at the University of Padua, where Gazzaniga, who collaborated with Ricci's occasional academic rival Giuseppe Veronese in the preparation of a geometry textbook for the secondary schools, also taught number theory, rational mechanics, and other topics.

Before deciding as a young teenager on a career in mathematics, Levi-Civita had evinced strong interest in classical languages as well as history, reflecting perhaps the influence of his one-time tutor Padrin, who also taught Greek and Latin for thirty years at the Liceo Tito Livio. Those interests faded as he grew older. Asked years later if Levi-Civita enjoyed reading different kinds of books, a family member replied, "I don't believe so. Or at least, only mathematics books," before adding, "Later, even though his father-in-law, Luigi Trevisani, had antique books—which I think he had bought himself, also in Latin—Levi-Civita bought only the volumes that he needed: he was absolutely not interested in necessarily collecting, as bibliophiles do, a complete set of works."[9] During his life, he did amass an impressive working collection of mathematics and physics books, well over 3,000 volumes, which his widow, Libera Trevisani Levi-Civita, donated to the Lincei Academy after World War II ended.

In his seven years as a secondary school student, Levi-Civita compiled an impressive attendance record, missing only one day of school. The story, as related to the author by historian Mariarosa Davi, the school's current archivist, concerns the last day of the school year, December 31. It fell on a Monday in the first semester of Levi-Civita's first year at the high school, and the school, which had been closed the previous week, reopened. Some students, including Levi-Civita, stayed home instead, choosing to complete the traditional vacation "bridge" between Christmas and New Year's Day. In doing so, he broke the rules and earned a 6 for conduct, almost half of his usual grade. "It was a slight deficiency, and very human, for such a good student,"[10] notes Davi. "And it makes him even more likable."

Unlike his future mentor Ricci, who continued to develop his absolute calculus throughout his career, Levi-Civita cast a penetrating spell on all of mathematics. "[He moved] among many fields, from one to another, without difficulty—from analytic mechanics to electromagnetism, from celestial mechanics to the theory of heat, from hydrodynamics to elasticity—and everywhere addressing fundamental problems characteristic of the way he thought about them,"[11] Amaldi, who knew him well, later wrote. Somewhat unimposing physically (barely five feet tall and very near-sighted, fearlessly observing the world through glasses with exceptionally thick lenses), Levi-Civita was a mathematical polymath whose work stands out for its quality, quantity, and range. Pigeon-holing Levi-Civita "as this or that kind of mathematician,"[12] the English algebraic geometer William Hodge later wrote, is pointless. One of the many mathematicians from abroad who spent time in Rome in the 1930s, Hodge later recalled being "particularly struck by the vivaciousness and precision of his discourse"[13] and his passionate interest in all sorts of scientific questions. He added, "Viewing his work as a whole, however, the dominating impression one receives is of an astounding command of the technicalities of pure mathematics, aided by an acute geometrical intuition, applied mainly to problems of applied mathematics."[14] Although several of Levi-Civita's early papers in the 1890s clearly fall within the realm of pure mathematics, by the time he and Ricci put the finishing touches on their definitive joint article on the absolute calculus in the closing months of 1899, there was probably no one else in Italy, aside from his teacher, who could match Levi-Civita's deep understanding and facility with the subject.

* * *

When Levi-Civita, flush with youthful academic triumphs, enrolled at the University of Padua in the fall of 1890, Gregorio Ricci, the professor with whom he would forge a remarkable professional bond and life-long friendship, was embroiled, yet again, in a protracted struggle to become a full professor. It had begun two years earlier when Italy's minister of public instruction, Paolo Boselli, had declined to act on a unanimous motion of Ricci's colleagues to promote him, because under the Casati Law, a series of provisions governing the nation's educational system, Padua's science faculty had reached its allowable limit of full professors. By the following year, however, two members of the science faculty—physicist Augusto Righi and chemist Giacomo Luigi Ciamician—had accepted positions at Bologna, meaning one full professor chair (Righi's) and one associate professor chair (Ciamician's) were now available. That December the science faculty *again* voted unanimously to promote Ricci—this time to the position left vacant by Righi's departure. There was no internal competition this time, but, perhaps inevitably, a contender emerged from left field. When Adolfo Bartoli, an experimental physicist then teaching at the University of Catania in Sicily learned about the vote, he mounted an aggressive campaign of verbal and written protests to fellow academics and various officials, asking that they open a *concorso* for Righi's replacement rather than simply awarding the post to Ricci. Naturally he had the ideal candidate in mind. "It would be a real disgrace," he lamented in a letter, "if the illustrious tradition of Padua's beautiful and richly furnished physics laboratory were entrusted to an inadequately trained scientific person."[15] According to the disgruntled Catanian's calculations, Ricci's promotion meant that "Padua would have eight full professors and only one associate professor in mathematics,"[16] which, he maintained, exacerbated the science faculty's already deplorable imbalance between the full professors in mathematics and those in the experimental sciences. Ricci, caught off guard by Bartoli's challenge and most likely experiencing an unwelcome sense of déjà vu (although Veronese, his victorious challenger in the previous go-round, was proving to be one of his warmest advocates in this one), vented his frustrations in a letter he wrote in January 1890 to Vito Volterra in Pisa. Recalling the fiasco of his first promotion attempt three years earlier and clearly dismayed at the prospect that history might repeat itself, he railed against Bartoli—"a relentless and willful person"—whose upstart meddling had become a constant thorn in his side, although he believed it also had alienated many of Padua's senior science faculty. That same day, he dispatched a second letter to Volterra, enclosing a third letter that he asked Volterra to pass on to his former teachers, Betti and Dini. What he actually wrote to them has not survived, but he likely expressed the same sentiments that he had shared with Volterra, particularly as Volterra reported that after reading it, Betti's advice was that Ricci focus on his own promotion and cease fretting about the activities of others.[17]

Matters continued to go downhill from there. That same month, the ministry of instruction's high council, apparently influenced by Bartoli's arguments, met to address the "strange disproportion" of full professors among the science faculty at Padua—more, they pointedly noted, than in most such university departments in Europe. While this did nothing to help Bartoli—Padua had by now considered and, as Ricci had more or less predicted, rejected him for its open full professorship—the council did adopt his view that the physics chair Ricci had hoped to occupy should instead be reserved for a suitable candidate in experimental sciences and recommended that the university throw it open to a *concorso*. If Padua was still bent on promoting Ricci, they suggested, this could perhaps be accomplished by transferring a full professor or two from the science faculty to the engineering school, thus keeping the number of full professors in the science department from reaching outlandish proportions. Needless to say, Domenico Turazza, the director of the engineering school, objected vehemently to this idea, as did the two hapless academics—one in ornamental design, the other in architecture—"volunteered" for the transfer. They refused to bow to outside pressure and insisted on staying where they were.

Ricci's continued frustration over this tangled state of affairs may be gleaned from a letter he received that February from his brother, Domenico, to whom Ricci had written about what he regarded as his ill-treatment.

Once again, his original letter has not survived, but in his reply Domenico wrote:

> You can't believe how distressed I felt to learn about your promotion [difficulties].... It is repugnant to me to dwell too much on how great the injustice is and on the ill will they must feel towards you. Certainly it can only be explained through hidden forces and influences.... What is there to say, my dear Gregorio? We need to face the fact that the world is much worse than we would have predicted when we were young. Nevertheless we must not lose heart, but... fight with courage and persistence until the very end. I'm only sorry that I cannot help you with this fight. You say that you have influential people on your side, and I trust they are so, but in these cases there can never be too much support. [Here Domenico offers to enlist a high-ranking local official on his brother's behalf.] Meanwhile, don't let disgust and displeasure burden you; stay calm and... remember that even if you lose the battle, it would not, thank God, have the same consequences for you that others might face in the same situation [evidently a reference to the fact that Ricci did at least have tenure].[18]

His gloom notwithstanding, Ricci was right to believe he had influential allies. In March, the Padua science faculty rejected the high council's recommendations, voting yet again to promote him immediately and to hold a national competition for either an associate professor of physics or chemistry. Soon after, his old teacher Dini, now a deputy in the Italian parliament, complained directly to Minister of Instruction Boselli in an emotional letter, in which he addressed him using the familiar "tu" instead of the formal "lei" to emphasize their personal and working relationship. After criticizing the council's seeming determination to designate the professorial chair in physics for the experimental sciences ("a very strange vote"), Dini continued, "It would be truly painful if yet again Ricci is left behind. I could explain this and related matters far better to you in person. For now I will confine myself to saying that I believe that in all of this *cronyism* and *nepotism* [italics in original] played and continue to play a formidable part."[19] Still, to all appearances the matter went nowhere until midsummer when Giovanni Garbieri, now the only other associate professor in Padua's mathematics department, left the university for a similar position at Genoa. Within weeks Padua announced a *concorso* for his position—that of associate professor of complementary algebra, which Ricci promptly applied for and won. In November, after years of dragging its feet, the ministry of instruction's high council approved his promotion to full professor of complementary algebra. Ricci's long struggle was finally over.

What had broken the impasse? First, it seems clear that Dini's letter to Minister Boselli had had the desired effect. The minister, who had studiously held himself aloof, either out of preference or policy, from the machinations of his council, suddenly reversed course and urged Padua to proceed with Ricci's promotion. Secondly, the council's influential vice president, Luigi Cremona, who had repeatedly balked at the prospect of promoting Ricci to full professor of physics, was much more willing to entertain the idea when the chair in question was one that had always been held by a mathematician. "Ricci's titles," he opined during a discussion of the matter in October 1890, "lie more in algebra than in mathematical physics," adding that "Ricci's value as a full professor in mathematical physics could be called into question, and I would want the judgment of a commission; but his record in algebra leaves no doubt as to [his] great scientific value... in that field."[20]

As for the surplus of full professors both in Padua's science department and among its mathematics faculty, here it appears that, to use a contemporary phrase, the fix was in, and that Ricci probably knew it before he agreed to what was essentially a lateral transfer into Garbieri's associate professor position. In a late fall report summarizing the events that led up to the council's approval of Ricci's appointment as *professore ordinario di algebra,* Dini alluded to a special dispensation authorized "under article 73," in Italy's educational codes, which allowed the council to override such strictures as the maximum allowable number of full professors within a given academic department. What the government bureaucracy had previously taken away, or at least withheld, it now granted—and Ricci got his promotion to full professor. One last faculty meeting held late that year at Padua confirmed his new appointment and recommended that although he no longer held the title, he continue teaching his course in mathematical physics (the course would remain his until his death in 1925). It was an upper-division course, taught in a two-year sequence. In 1892, Tullio Levi-Civita enrolled in the class.

"[I] had the good fortune and honor to be his favorite disciple,"[21] is how Levi-Civita, decades later, would describe his rapport with the professor he likely encountered for the first time in the fall of 1890 when, as an entering freshman, he took Ricci's course in complementary algebra. There is no record of precisely when teacher and pupil—the patrician, reserved mathematician who hailed from devoutly Catholic gentry and the brilliant, gregarious student whose Jewish forebears had made the leap from the ghetto to prosperity and political prominence in barely a generation—first began to recognize their intellectual affinity; what is clear is that once they did, they forged a connection that would endure for the balance of Ricci's life. Already in that first algebra class, Levi-Civita was struck by the caliber and clarity of Ricci's lectures, writing thirty-five years later, "This course... which I attended when he gave it for the first time, has forever remained for me a model of impeccable reasoning and fruitful mathematical enterprise."[22]

Ricci's class was one of many that Levi-Civita had to take as a requirement for the laurea in pure mathematics. The courses were like mother's milk—he kept going back for more. He learned higher mechanics and rational mechanics from Ernesto Padova and infinitesimal analysis and higher analysis from Francesco D'Arcais. Theoretical geodesy he mastered from lecturer Francesco Miari-Fulcis; descriptive and projective geometry he picked up from Enrico Nestore Legnazzi The director of the Astronomical Observatory of Padua grounded him in astronomy, using the observatory as his classroom. He found time for electives too: from classes in mathematical exercises to the theory of numbers, a subject begun in high school under the direction of Paolo Gazzaniga, who also taught at the university.[23]

In 1892, at the start of his third year, Levi-Civita enrolled in Padua's teacher-training program in mathematics, the prerequisite to an academic career at either the high school or university level. Giuseppe Veronese, who was then in charge of the mathematics section's geometry course, lectured that year on infinity and infinitesimals. He had recently produced a king-sized book on multi-dimensional geometry[24] whose methodology, using infinitely large and small numbers, had embroiled the combative professor in yet another controversy—this one centering on the notion of infinity in mathematics. Today regarded as a pioneering if somewhat uneven work that opened the door to the possibility of a non-archimedean geometry, Veronese's book, required reading though it may have been at Padua, was not universally appreciated. In particular, it drew the scorn of influential German mathematicians, who regarded it as unwieldy and synthetic. In the words of a modern German historian of mathematics, Detlef Laugwitz, Veronese's presentation "lacked clarity, and some of his formulas were obviously nonsense when one did not read the accompanying text with care."[25]

Despite these attributes, or perhaps because of them, the book inspired Levi-Civita's very first research paper,[26] "On infinities and infinitesimals as analytical elements," which he submitted for publication in May 1893, one year before he graduated. In his biographical memoir of Levi-Civita prepared in 1946, Amaldi remarks that this paper, which established the first rigorous construction of non-archimedean ordered fields, "seems even today the mature work of a proven researcher rather than the first work of an eighteen-year old."[27] (Levi-Civita in fact turned twenty that year.) Five years and seventeen papers later on a wide range of topics in analysis, mathematical physics, and mechanics, Levi-Civita would return one last time to Veronese's geometry with a work on transfinite numbers, in an effort to put to rest "misunderstandings"[28] by prominent critics of his former teacher's geometrical hypotheses.

Nevertheless, by his senior year at Padua (1893–1894), Levi-Civita had succumbed to the lure of Ricci's absolute differential calculus. While he followed Veronese's lectures on irrational numbers, took a full complement of courses, and busied himself turning out papers on assigned topics in analysis, geometry, and mechanics,[29] he also began work on his dissertation, choosing a topic that, as he notes in the first footnote of his thesis, Ricci had suggested to him. "On differential invariants"[30] marked Levi-Civita's first contribution to the absolute differential calculus. Inspired in part by Ricci's covariant differentiation and in part by Sophus Lie's theory of groups of transformations, he extended the theory of differential invariants to more general cases than those considered by his professor. Working on the absolute calculus brought Tullio into frequent contact with its inventor, who, as their subsequent relationship makes clear, had rarely come across a student whose aptitude for study and mathematical research, taste for complicated calculations, and breadth of interests, set him apart from his peers. It must also have given Ricci considerable satisfaction that his mathematical formulations, whose relevance and utility had been questioned and even disparaged at times, had captured the interest of this exceptionally gifted student. For his part, Levi-Civita, as he would attest many years later, found in Ricci a truly admirable mentor, whose "unflinching moral rectitude, natural reserve, and the serenity with which he accepted the judgments of others equaled the power of his intellect."[31]

In July, after completing a dozen special exams with a perfect score of 30 in nearly every subject (*con pieni voti assoluti*),[32] Levi-Civita successfully defended his degree in mathematics. Soon after, the much-praised and hard-working *Dottore in Matematica con lode* ("with honors") boarded the train for Bologna, where a new generation of mathematicians trained at the Scuola Normale in Pisa had ushered in a new era of mathematical teaching and research at the university. He remained there for about six months, long enough to become more familiar with Salvatore Pincherle's research on functional operators and to strike up a life-long friendship with twenty-three year old Federigo Enriques, an instructor in projective and descriptive geometry. Writing to his colleague Guido Castelnuovo in Rome, with whom he had begun a collaboration on the geometry of algebraic surfaces, Enriques described his new acquaintance as "a talented young man and very studious, also nice personally; and for me a pleasure to be able to have a mathematical conversation with... although rarely on geometrical arguments."[33]

Returning to Padua in spring 1895, Levi-Civita was quickly pressed into service as an assistant not only to Ricci in algebra but also to D'Arcais in infinitesimal calculus and Veronese in geometry. It seemed he would not remain there long since the following year he applied for the position of *libera docenza*—essentially an instructor—in infinitesimal calculus at the University of Pavia.[34] The Pavia commission that provisionally approved his appointment expressed some skepticism about his early paper on absolute invariants, describing the first part (a presentation of Ricci's methods) "as containing complicated calculations perhaps irreproachable, but certainly not elegant, and the second part (the problem of invariant integrals) as "research that is essentially only a different form of research already noted." They were far more impressed with his more recent publications on functional analysis from the point of view of group theory, which they considered a sign of "the author's uncommon insight and analytical ability."[35] Pavia's science faculty seconded this judgment and sent it on to the ministry of instruction's high council for approval. Only a week later, however, Ernesto Padova, professor of rational mechanics at Padua and Ricci's closest colleague in the department, died after a lingering illness. Without a moment to waste, and with no mention of his impending appointment at Pavia, Levi-Civita requested Minister of Public Instruction Luca Gianturco's assistance in obtaining the rank of *libera docenza* in rational mechanics at Padua.

Although Padua's science faculty may have dithered and delayed over Ricci's promotions, it moved with lightning speed in Levi-Civita's case. On May 16, the commission appointed by Padua's science faculty to referee Levi-Civita's publications—Ricci, D'Arcais, and Volterra—prepared a detailed report on the candidate's works. Unlike the examiners in Pavia who had dismissed Levi-Civita's approach to Ricci's calculus out of hand, this trio of reviewers merely sounded a cautious note. "Following Lie, Levi-Civita chose to study the differential invariants shared by multiple tensors, without a requirement that these include the metric tensor,"[36] they wrote. "One can perhaps argue," they added, "about the appropriateness of the choice of such an argument, which reveals moreover a tendency that appears in other works of the author, to prefer general theories in comparison to applications to concrete problems." Having surveyed the entirety of his works, the commission had only praise for the breadth of his knowledge, the exceptional ingenuity of his calculations, and the stratagems that he employed in pursuing a wide range of problems. Padua's rector then forwarded their endorsement directly to Minister Gianturco. If he was attempting to fast-track the appointment by making an end-run around the high council, whose approval was usually solicited first, he failed. The minister promptly asked his council for its assessment of Levi-Civita's qualifications. The council responded by inquiring whether the brash young petitioner intended to secure positions at two different universities simultaneously. Shortly thereafter, in early October, Levi-Civita formally notified the minister that he had withdrawn his request at Pavia, leaving only Padua's application active.

With that issue clarified, the high council gave its blessing to the appointment, singling out in particular the favorable attention that Levi-Civita's recent papers in mathematical physics and mechanics had received from scientists abroad. That November, eager to see Levi-Civita's appointment take effect as quickly as possible, university officials at Padua arranged for Eugenio Valli, a deputy in Parliament and evidently their "inside man," to speak to a colleague in the minister's office. Valli followed up with a short reminder note several days later, stating pointedly, "I would greatly appreciate if the honorable minister would *sign* the official decree immediately."[37] In December 1896, Levi-Civita received official news of his appointment. A mere two years later, in 1898, he entered the *concorso* to fill Padova's vacated chair at the associate professor level.

Although the competition began with four candidates vying for the open position, it ultimately pitted Levi-Civita against Roberto Marcolongo, a dozen years his senior and already an associate professor at the University of Messina (Sicily again! one can imagine Ricci saying in exasperation). Possibly swayed by the prerogatives of seniority, the Padua judging panel split 3-2 in favor of Marcolongo, but in giving him the green light added "that in case Professor Marcolongo does not intend to hold the position, Doctor Levi-Civita could be nominated without need of further examination."[38] This may have reflected the views of the minority commissioners, one of whom, Vito Volterra, was not shy about later stating that he had a much higher opinion of Levi-Civita. (He had previously written to a colleague that "Levi-Civita is a young man of great creativity and uncommon worth, and he is one of the most beautifully promising among young Italian mathematicians."[39]) In the end, it was Volterra's judgment that prevailed. Early in 1898, Minister Gianturco informed university officials at Padua that "to provide for the chair of rational mechanics that remains vacant [at Padua], I have nominated [as] associate professor Doctor Tullio Levi-Civita... effective, January 16 of this year."[40] Within a year Marcolongo had been promoted to full professor at Messina. In 1902 Levi-Civita, then twenty-nine, became a full professor of rational mechanics at Padua. He continued to teach there until 1918, when he went to the University of Rome as professor of higher analysis.

There is no indication that Ricci ever resented or was troubled by the fact that his former student's academic rise was both swifter and smoother than his own; in fact, as their intellectual collaboration grew, the opposite was probably true. Colleagues for more than twenty years, the absolute differential calculus was the catalyst that launched Ricci and Levi-Civita's scientific discussions, paved the way for their seminal 1900 paper on Ricci's absolute calculus and cemented a deep and enduring friendship.

Ricci's self-described disciple was not shy about invoking his mentor's methods in his research. Turning to the field of analytic mechanics in 1896, Levi-Civita published "On the transformation of dynamical systems," the first of a series of memoirs concerning the study of dynamical equations. A number of French mathematicians had already worked on the general problem of the equivalence under transformations of two systems of dynamical equations without reaching any definite conclusions. In his summary of this paper later, Volterra notes that by limiting himself to the case in the absence of an applied force, Levi-Civita transforms the original problem into a geometric one. He "was thus led to the related geometrical problem of the representation of a Riemannian manifold of any dimension by way of geometric congruences, which reduces the question to the study of the variety to which the geodesics correspond, and he distinguishes these varieties to as many types as the number of their dimensions. To each type corresponds a certain number of distinct quadratic integrals of the dynamical equations that he completes with an important theorem due to [Roger] Liouville."[41]

While Levi-Civita's explicit use of Ricci's absolute calculus figured prominently in his article, he also pointed out that Ricci's methods "up to now have not become as widely used as perhaps might have been desired."[42] Unlike Volterra, who avoided the subject, Amaldi emphasized its role in his own summary of Levi-Civita's paper: "In [his] work, the absolute differential calculus that until then Ricci—perhaps also hindered by the incomprehension of most mathematicians at that time—had used almost exclusively within the traditional boundaries of metric differential geometry, was for the first time able to demonstrate its power in the treatment of a new and interesting problem, in comparison to futile results that would have resulted from less penetrating investigations."[43]

Less demonstrative by nature than his younger colleague, Ricci never offered a public testimonial to their collaborations on the order of Levi-Civita's tribute following Ricci's death in 1925. But a letter that he wrote as Levi-Civita prepared to leave Padua for the University of Rome in 1918 leaves no doubt as to how much their personal and professional relationship meant to him:

> I don't need to repeat my feelings, which you already know, and could certainly also guess, knowing the real and profound affection I have for you and the esteem I have for your research and teaching. Since it is only human that selfish feelings will overcome altruistic ones, you may easily guess that the sorrow I feel for your distance from Padua, and for the end of our personal and professional interactions, almost outweighs the pleasure that I must feel to see one of your fondest wishes [i.e., the academic chair at Rome] granted. I cannot close this without assuring you that I was greatly moved by the extremely gracious sentiments expressed toward me that I found in your letter and for which I am very grateful.... Please remember me to your kind wife [Levi-Civita had married in 1914] and accept this expression of my unshakeable affectionate friendship for you.[44]

CHAPTER 9

# Intermezzo

Tullio Levi-Civita knew Gregorio Ricci for more than a quarter of a century. He probably possessed greater insight into his personality and temperament than any other mathematician of Ricci's generation, which helps explain why the commemoration that he delivered at the Academy of the Lincei in 1926 is an essential resource for anyone studying Ricci's life. Drawing on his varied experiences as student, colleague, collaborator, and friend, Levi-Civita presented a discreet and respectful portrait of Ricci's lengthy struggle to gain proper recognition for his absolute calculus, the obstacles he encountered in his quest to become a full professor, and his lack of success in garnering the prestigious prizes that cemented the reputation of several of his contemporaries. He described Ricci's reactions to these setbacks as evidence of his strength of character—serene in the face of repeated disappointment, reserved and dignified by nature; in short, a man of calm, austere fortitude, determined to pursue and perfect his mathematical inventions and their applications in the face of general indifference from the Italian mathematical community.

Nevertheless, there is evidence, in both his private correspondence and in the public record that in the 1890s, Ricci underwent what we might call today a midlife crisis, a situation that is barely alluded to in Levi-Civita's sympathetic portrayal. Throughout a good deal of the decade, Ricci appears to have been a man out of sorts, nursing perceived grievances, quarreling with colleagues, Levi-Civita included, and heaping guilt on his beloved wife when she seems to have felt that some time spent apart might benefit at least one of them.

A mathematician who shunned professional conferences and regarded presenting papers at them as an ordeal to be avoided whenever possible, Ricci was much happier spending summers with his wife, Bianca, and their children in his ancestral village of Lugo and the surrounding countryside. There, in the nearby village of Sant'Agata sul Santerno, he had purchased a large eighteenth-century country house located just outside the town's ancient walls. In his spacious book-lined study, reached by mounting a grand staircase from the first floor of the villa, he was able to work undisturbed while enjoying the view of his orchard and vineyards, which were faithfully maintained by a local farmer and his family.

In the fall of 1894, Ricci and his three children—Livia, Cesare, and Giorgio—packed up and returned to Padua. He seems to have expected that

Bianca would join them almost immediately; but having arrived home with his charges, who had, as he reported in a letter, stuffed themselves with prosciutto on the train ride and were now suffering the almost inevitable digestive consequences, he received a postcard from his wife. It contained the unwelcome news that she would be visiting relatives and friends in her hometown of Imola before rejoining the family in the city some days later. As Ricci quickly discovered, coping single-handedly with three youngsters under the age of eight left a great deal to be desired. Cesare, after spending the first day at home asking incessantly for Bianca, promptly transferred his attention to his papa, who described the experience as "a little annoying," possibly because it meant he had no time to get any work done in his home office. Still he found time to write to his wife before she left Lugo, instructing her to have one of the servants ship bottles of freshly canned tomatoes to Padua and reminding her to collect all the bed linens and arrange for them to be sent home.[1]

Five days and a number of letters later, and having heard nothing beyond another postcard from Bianca, Ricci unleashed a flood of reproachful words:

> My darling,
> Despite having repeatedly written to you, I remain today again without a letter from you. I am bitterly disappointed that since we have been separated, you have not written to me at length. Why are you so stingy with your letters, whereas I, as soon as I knew you desired it, have spared no detail in writing to you? I wrote long and expansive letters to you. You cannot be so busy that you cannot devote a little of your time to me. I want to hope that the impediment does not spring from a physical ailment; but really, what am I left to think? Am I to believe that your old and proven affection for me has worn itself out? Seeing as we [i.e., the family] have never been this far from you for as long as a week and having received only two paltry postcards, I prefer to believe that you have merely become more indolent. Indeed, why would you not love your darling like you once did, while he loves you more than ever and shows you that, I think, in every possible way? I repeat: I don't want to believe it, but I cannot completely kill such nagging doubts, and so I insist that you relieve me of this pain immediately and tell me truly the reasons for your unusual conduct. Out of spite and to test you I was almost tempted not to write to you and wait for you to demonstrate that you are alive; but then I realized that would not do me any good. But now you are warned, and if you don't immediately write me a good, long letter, you will not hear anything else from me![2]

While offering glimpses of Signore and Signora Ricci's domestic life (handling bills and bank deposits, dealing with the household help, choosing schools for the children), this missive ended on a decidedly barbed note. "This time you have really gone too far," her husband admonished Bianca. To her family back in Imola, he dispensed the customary hugs and kisses. To his wife he offered "a good spanking!"

Bianca's two-part peace offering arrived swiftly: an appropriately conciliatory letter and a fine photograph of herself, prompting Ricci to reply, "I take back the spanking with all my heart and in exchange give you a million kisses."[3] The minor domestic crisis was over. But smoothing over professional slights took a far greater toll on Ricci's customary imperturbability. It has been observed that few scientists, no matter how successful, ever feel that their work has been fully appreciated.[4] Ricci at this time of his life was no exception. He assumed, perhaps naively, that if the right problem using the methods of his absolute calculus came along, appropriate acknowledgment and recognition would inevitably follow.

It was certainly not for lack of trying. Three times, Ricci entered contests for mathematics prizes, and each time, a different set of circumstances derailed his chances for winning. If Ricci was disappointed when the judges chose not to award Italy's first Royal Mathematics Prize to him (or anyone else) in 1887, he remained stoic. His commitment to his work was unaffected, according to Levi-Civita, who later commented that "while *a priori* confidence should never influence judgment, it never failed to spur Ricci's research."[5]

Unlike the Lincei's single prize in mathematics, the French Academy of Sciences offered a relative bounty of awards in various mathematical fields, including the prestigious Prix Bordin in pure mathematics. In 1888 the celebrated Russian mathematician Sofia Kovalevskaya had won the prize with a paper on her use of the theory of Abelian functions and hyperelliptic integrals to solve a physics problem. That same year the academy announced that its next prize would be awarded for work in differential geometry, a subject avidly pursued by French mathematicians.[6] Although it is not clear when Ricci first got wind of this competition, in 1892 he finished writing up his solution of a problem concerning geodesic lines and entrusted someone (perhaps his wife or a servant) to post the envelope containing his paper to Paris.

As the months passed with no response or acknowledgment from the academy, Ricci wrote to Eugenio Beltrami, who was perhaps Italy's leading differential geometer, asking him to make some discreet inquiries about the status of his paper. Beltrami in turn immediately dispatched a letter to a member of the judging panel and his counterpart at the Sorbonne, the differential geometer Jean Gaston Darboux. As Beltrami later wrote to Ricci, Darboux seemed inclined to dance around his inquiry but finally admitted that the article in question might have gotten lost or perhaps "confused" with another submission. He himself, he assured Beltrami, had not yet had a chance to review the papers, although he knew that since announcing the contest four years earlier the commission had received a total of four entries, including one from Italy. When Beltrami wrote again, requesting more details, the Parisian mathematician replied, according to a letter Beltrami then sent Ricci, "that he knew only this, that '[the author] makes frequent references to the work of Mr. Ricci.'"[7] Ricci's reaction when he read this can only be imagined. Beltrami's advice, that in the future Ricci would do well to mail packages abroad himself and to ask for a return receipt, presumably offered small consolation. Shortly thereafter, the Prix Bordin was awarded to none other than one of Darboux's former students, the well-known French analyst and geometer Gabriel Koenigs.

A near contemporary of Ricci's (he was born in 1858), Koenigs had his eye on the prize, in every sense of the word. Raised in Toulouse, he had studied under Darboux at the École Normale Supérieure in Paris, where he would return to lecture in mathematics. He also held an appointment at the Collège de France in the capital, where he taught analytical mechanics, devoting the 1892–1893 academic year to a yearlong course in geodesic lines. By his own account, he had published two papers on the subject in the *Comptes Rendus* [*Formal Report*], in 1889. In his annotated bibliography, which appeared in 1897, Koenigs recalled that after submitting these two works to the academy in 1892, the judges declared him the winner of the Prix Bordin.[8] That December, he published a summary of a memoir on geodesic lines in the *Annals* [*Annali*] *of the Toulouse Science Faculty,* where Ricci subsequently spotted it,[9] several months after he had presented the results of his own research on geodesic lines at the Lincei Academy, in January, 1893.[10] Reading Koenigs's summary, Ricci learned for the first time from a helpfully supplied footnote that the Frenchman had won the Prix Bordin in differential geometry.

The Bordin prize was not without its problems. As mathematics historian Jeremy Gray has pointed out, in several notable cases, including that of Kovalevskaya, the theme seemed to have been chosen with a specific candidate's research in mind. In general, adds Gray, a number of more problematic episodes regarding the selection process suggests "that all these topics were set with a shrewd eye to who was working actively on what subject."[11] Although Gray does not press this point in Koenigs's case, the letter Beltrami wrote to Ricci would seem to fit the pattern: Darboux and his fellow committee members had prejudged the matter before reading the submissions. Of the four applicants for the prize, honorable mentions went to the other two entries (one from Germany, one from France), leaving the "lost" Italian entry to fade away, unnoticed.

In September of 1893, after quietly stewing about the French Academy affair for some months, Ricci published a brief note in the Lincei's *Proceedings,*[12] indicating where his results and Koenigs's coincided and where their methods parted company. In his reply, which appeared in the same issue,[13] Koenigs remarked that his methods gave a more general result, and he referred readers to a recent article of his in the *Comptes Rendus*. Unbowed, Ricci responded with his own rejoinder, insisting that his results completed the Frenchman's theory, before getting down to the root cause of their disagreement. "When I became interested in this topic, which the Paris Academy of Sciences had repeatedly called to the attention of geometers, I intended to show the effectiveness of the methods of the absolute differential calculus."[14] He continued,

> This effectiveness derives from the fact that [my] methods eliminate in a natural way all irrelevant elements from problems that are independent of the choice of the coordinates. Indeed, I hope to have given a convincing example of their effectiveness, since if I am not mistaken, the problem that I solved would have been undertaken by ordinary differential calculus with difficulty. Finally, I will say that I became aware of Koenigs's summary[15] only after the publication of my Note in the January 22, 1893 issue of this *Proceedings*.

In 1894, Ricci published a greatly expanded work on the theory of geodesic lines ("On the theory of geodesic lines and isothermal Liouville systems"), partly to drive home the superiority of his absolute calculus, and partly to remind his readers that most of the results presented in this paper had been published before he read the summary of what he generously termed his French colleague's "beautiful" memoir. While the Paris Academy of Sciences, he noted, had "crowned" Koenigs's memoir as a *singular* achievement, in fact their results had much in common, despite being reached "not only by different paths, but independently from each other."[16] Left unsaid, but definitely on Ricci's mind was the fact that Koenigs had simply ignored his detailed, French-language account of the absolute calculus, published in the June 1892 issue of Darboux and Paul Tannery's influential *Bulletin des Sciences Mathématiques*—a journal explicitly founded with the aim of expanding French awareness of foreign mathematical research. In Ricci's eyes, Koenigs had essentially failed to live up to the French journal's own goals. After this brief spate of notes and counter notes, the dispute seemed over, and Koenigs's prize-winning memoir was finally published in 1894.[17] Buoyed by the Prix Bordin and two additional prizes, Koenigs's academic career reached new heights: In 1895, he left the Collège de France for a position as assistant professor of physical and experimental mechanics at the Sorbonne, which promoted him to full professor two years later. Ties with Darboux, his mentor, remained strong, and although his research shifted dramatically to new ways of teaching mechanics and developing an applied mechanics laboratory, he also contributed a lengthy appendix on geodesics to the last volume of Darboux's classic work on the general theory of surfaces. Not surprisingly, there are no references to any of Ricci's papers on the same subject.

Before choosing a name for his calculus in 1893, Ricci had made a point of selecting various problems in analysis, physics, and geometry as vehicles for demonstrating the utility of his novel methods and procedures. Afterward, he increasingly focused his absolute calculus research program on problems that could be studied from a geometrical perspective, and in the process, began reinventing himself as a differential geometer. Nevertheless, analysis, that branch of mathematics that had developed from the calculus of Newton and Leibniz, still claimed his intermittent attention, along with his penchant for vigorously defending his point of view. In the early months of 1895 he entered into a brief but spirited correspondence with the University of Naples algebraic analyst Alfredo Capelli about number theory. Significantly, in a sign of his confidence in Levi-Civita's judgment, he also asked his young colleague to weigh in on the argument, and he incorporated Levi-Civita's thinking before sending Capelli his final text. Utilizing his considerable proficiency in number theory, Levi-Civita embarked on a detailed analysis of Capelli's side of the correspondence, invoking the names and ideas of mathematicians from Euclid to Dedekind.[18] Summing up his report, Levi-Civita wrote, "In my opinion, I would add the observation that the demonstration of the fundamental theorem of limits, which is found in Professor Capelli's lessons [referring to his 1895 book, *Lezioni di Algebra Complementare*], is much more artificial than that of Dedekind's. The same can be said... for the fundamental theorem of algebra."[19] In his later years, Ricci described this exchange with Capelli as "a controversy," but conceded it had been a fair argument on both sides.[20] His season of discontent seemed to have ended.

In 1896, however, Ricci's frustration over the Koenigs matter flared again. Ironically, the catalyst was Levi-Civita, then winning golden opinions for himself during his year at Pavia. Having drafted his most important paper to date, "On the transformations of dynamical systems,"[21] outlining the conditions under which two systems of dynamical equations can be transformed into one another, he wrote that spring to Vito Volterra about it, proudly noting, "In part it is only a refashioning of results obtained already by [Paul] Appell, [Paul] Painlevé, and [Roger] Liouville, but having adopted a different method and, in my opinion, better responding to the nature of the question, it spontaneously presented me with certain consequences, which they had overlooked, at least in the case where forces do not act, to the complete resolution of the problem."[22] That "different method," involved applying and extending aspects of Ricci's absolute calculus for perhaps the first time in the context of this particular problem, and Levi-Civita probably anticipated a warm reception, to say the least, when he mailed a draft text for review to his former professor in Padua. Instead his unfailingly supportive mentor took violent exception to a footnote.

Ricci read Levi-Civita's text in late June and almost immediately communicated his displeasure with its references to Koenigs's work on geodesic lines. In particular he objected to his young colleague's treatment both of

Koenigs's summary paper in the Toulouse *Annali* and his own geodesics paper presented to the Lincei Academy in January 1893. The sticking point was a footnote that he and Levi-Civita had apparently already agreed to insert in the memoir, and which Ricci pointedly remarked did not give his geodesics paper—and more specifically his results—sufficient credit. As Ricci wrote in the first of three letters dispatched to his young colleague within the space of one week,

> It seems to me that the wording of your Note, which you suggest be inserted in your memoir, does not correspond entirely with the actual state of affairs. In order [for you] to follow more easily the points that I think should be changed, it is perhaps advisable for me to actually explain how I see things, cautioning that while I am sure about everything that concerns me, there could be some misunderstanding concerning Koenigs's work, as I did not see his extensive memoir devoted to this theory.[23]

In his 1892 paper, continued Ricci, he had laid out the criteria for identifying whether a given linear element is reducible to Liouville's form, followed by a demonstration of allowable isothermal Liouville systems. "In my opinion," he told Levi-Civita, "it is not correct to say that it is necessary to resort to Koenigs's methods to determine all Liouville systems for which a certain linear element is provided. My methods are instead suitable for doing this directly." He then turned to the matter of whose results deserved mention in Levi-Civita's paper: "It seems to me... that only my results can be applied and not [Koenigs's], unless, I repeat, his more extensive memoir contains results not mentioned in the summary."[24] His aggrieved tone notwithstanding, Ricci concluded the letter by basically handing the problem back to Levi-Civita. Handle the footnote as you think best, he told him: leave it in or omit it, and let the reader follow the thread of the argument by himself, if he so chooses.

Although we lack Levi-Civita's reply, it can be surmised from Ricci's quick response, four days later, that Levi-Civita's amended citation had not yet mollified his feelings. "I have to thank you for the time you are devoting to a question that mainly interests me alone,"[25] Ricci's letter began. He then plunged into an animated discussion of why

> Koenigs did not provide any contribution to the solution of the problem, which concerns the effective reduction of a given element, if possible, to Liouville form. In that case, you will find it just to leave the credit for this [solution] to me, just as the [Bordin] prize fell entirely to him, and what is more, the honor attached to it. Having said all that, I do not mean to dictate the words to substitute for those that concern this point in your manuscript; all the more so as I would not like to put you in a position of

> having to defend me in a polemic against Koenigs. I really don't relish this kind of controversy, not least for the loss of time; and therefore I would not like you to suffer on my account. If what I have written has convinced you, then do what you think best.

By now, Levi-Civita must have understood what was really gnawing at his former professor and why he kept harping on a "question" that he acknowledged was of no interest to anyone except him. By return mail, he assured Ricci that he now saw "things in the same way"[26] and that he would certainly insert an appropriately worded footnote in his paper—which he did.

Toward the end of his 45-page memoir, which in its final published version omitted all mention of Koenigs's paper in the body of the text, Levi-Civita observed, that for "$n = 2$ [where $n$ is the number of independent quadratic integrals], the problem has been completely solved by Prof. Ricci," and in the promised footnote he summarized the arguments Ricci had himself put forward. However, Levi-Civita was determined to have the last word, and in the same footnote he wrote, "It will not seem strange that I did not mention the important and fundamental memoir of Koenigs on geodesic lines, when one considers that in all his investigations he substantially assumes the linear element is already reduced to Liouville form, and only then examines its most hidden characteristics and establishes its properties, which while remarkable, are not connected to the problem at hand."[27]

It had taken some nimble footwork on Levi-Civita's part to defuse the situation and restore Ricci's equanimity, but as the last line of the footnote reveals, Ricci's protégé managed to carry his point as well. Indeed, in his acknowledgement of Koenigs's work, Levi-Civita displayed the independence that would be a hallmark of both his academic and his public life. Whether the French Academy played by the rules in awarding Koenigs the Prix Bordin in 1892 remains an open question. Once his memoir with its much-debated footnote had been published in the *Annali di Matematica pura e applicata,* Italy's premier mathematical journal, Levi-Civita sent copies to a number of French mathematicians, including Koenigs, who never acknowledged receiving it. "Several other French [mathematicians], however, were most considerate; [Roger] Liouville, for example, sent me a note and several of his own reprints,"[28] he later informed Ricci.

In his 1926 biographical memoir of Ricci, Levi-Civita avoided all mention of the Koenigs affair, writing only that "An example [of showing the applicability of the absolute calculus] was [Ricci's] intrinsic characterization of surfaces for which $ds^2$ was reducible to Liouville's quadratic form."[29] His contemporaries and Ricci's, assuming they had heard anything about it, probably regarded it as much ado about nothing. Its most lasting impact may have been to give Ricci and Levi-Civita a better appreciation of each other's characters and greater confidence in their working relationship, setting the stage for a collaboration that would immeasurably enrich twentieth century science.

* * *

In August 1897, Levi-Civita, now permanently back at Padua and teaching the late Ernesto Padova's course in rational mechanics, joined Ricci, Volterra, and hundreds of other participants from sixteen countries at a three-day meeting in Zurich billed as the first International Congress of Mathematicians. For Ricci, it offered a chance to cross paths again with Felix Klein, his former professor in Munich and one of the scheduled speakers at the conference. During the meeting, Klein, a longstanding member of the editorial troika of *Mathematische Annalen,* the world's premier mathematical journal, invited Ricci to write a review article about the absolute calculus for the journal's readers. It was apparently an open invitation, as Ricci and Levi-Civita only began work on the piece two years later. Then nearing fifty, Ricci had never before written anything with a colleague, nor would he in the future. It is not entirely clear how Levi-Civita came to be a coauthor on this article; Klein may have extended the invitation to him as well, or the two mathematicians may have decided on their own to work on it together. The historian can only speculate, as the record is silent.

Whatever the impetus, they were making good progress when Klein visited Padua in March 1899. In a letter that month to the German physicist Arnold Sommerfeld, Levi-Civita mentioned Klein's "very welcome visit,"[30] along with the news that he had "been busy writing" an article for the *Annalen* "on the methods of the absolute differential calculus of Prof. Ricci and on several applications." He added, "I have had the opportunity to draw on some of these [applications] myself." By late autumn, the two mathematicians had completed the writing, save for putting the final touches on the report, which "for me," Levi-Civita would privately admit, had been "extremely tiring."[31] Written in French, Ricci and Levi-Civita's memoir, titled simply "Methods of absolute differential calculus and their applications,"[32] was only translated into English in 1975 by the American mathematician Robert Hermann, who described the paper "as one of the most influential and important in the history of *both* differential geometry and mathematical physics."[33]

As early as 1884, Ricci had taken other mathematicians to task for not using transparent methods in their work. Delighted to find similar sentiments expressed by an illustrious colleague in France, he and Levi-Civita opened their joint article work with their own musings on what constitutes a mathematical proof: "M. Poincaré has written that in the mathematical sciences a *good notation has the same philosophical importance as a good classification in the natural sciences* [italics in original]."[34] That statement was followed by a second one, equally emphatic in tone: "Evidently, and even more so, one can say, according to Poincaré, the same thing about methods, since it is certainly their choice that creates the possibility *to group together a multitude of facts, which have no apparent relation, according to their natural affinities* [italics in original]." In addition to advocating the use of proper methods to solve problems, the authors addressed another pet peeve of Ricci's—and one likely to have been imparted to Levi-Civita by his mentor—this one dealing with the matter of precisely what constituted mathematical proofs. "One can also say," they wrote, "that a proof of a theorem utilizing mainly artifices and considerations that have no essential relation to the theorem is often really only a partial truth; in fact, it almost always is the case that the same theorem is understood in a more complete and general way if its proof is more straightforward and uses more appropriate means." Conceived as a general account of the whole subject, their brief exposition of the theory and methods of calculating with the absolute differential calculus came wrapped in an ambitious goal: to recruit a generation of scientists willing to expend the effort to both learn the new techniques and use them in their research.

After presenting the mathematical terms and tools used in the calculus, Ricci and Levi-Civita applied them to a number of problems in differential geometry, mechanics, and physics in order to demonstrate the heuristic value of their "good notation." Indeed, what is striking is the number of works cited in the report (and in Ricci's case, also textbooks) that both mathematicians had already written using the machinery of the absolute calculus.[35] By December 1899, the paper was finished and dispatched to the *Annalen* (presumably accompanied by a return receipt). It would be published the following year.

In many respects 1899 was a propitious year for both mathematicians. Levi-Civita had recently been appointed associate professor of rational mechanics at Padua. Ricci had been elected a corresponding member of the Lincei Academy, a signal honor (and one of his first), and was reviewing for one last time his classroom lectures on topics in algebra before handing them over to the publisher to be lithographed. He had already completed a series of studies analyzing problems related to the theory of surfaces using the methods of his calculus, which he then expanded into a complete account on the subject, *Lessons* [*Lezioni*] *on the theory of surfaces.* Issued the year before as a lithographed volume by the university's in-house printer and bookshop, [36] *Lezioni* was enthusiastically received in America. Reviewing it in the *Bulletin of the American Mathematical Society,* George Oscar James, a recent graduate of Johns Hopkins University, where his dissertation had focused on differential equations connected with hypersurfaces, wrote: "The instrument of analysis termed Calcolo differenziale assoluto by the author,"[37]

> leads to formulae and equations always presenting themselves under the same form for any system of independent

> variables, and the difficulties which are incidental and formal rather than intrinsic are thus to some extent done away with, and the research assumes a uniformity absent in other methods. The entire discussion is based on the properties of differential quadratic forms, and the Introduction... contains a rather complete exposition of the methods of the calculus with a more or less full treatment of differential quadratic forms in general and binary forms in particular.

He admitted that "the novelty and generality of the method and notation" demanded some effort on the part of the reader, but if that effort was made, "the various geometrical applications follow easily." The idea behind the work "seems to be to give some notion of the power and elegance of the absolute calculus"—which indeed had been the driving force behind Ricci's research from the start. Mathematicians "accustomed to the classic method and notation, as found in Darboux and Bianchi for instance, will find the book well worth reading if for no reason, than to look at the theory from another and entirely different aspect," James concluded. It was in fact a warmer endorsement than Ricci had often received in Italy.

For years, Ricci had tried to interest the University of Pisa's renowned differential geometer Luigi Bianchi, a former classmate of his at the Scuola Normale, and the first winner (with Salvatore Pincherle) of the Lincei Academy's Royal Prize in Mathematics, in reading some of his papers on the absolute calculus. It was, at all times, an almost painfully restrained series of overtures. In letters he wrote to Volterra, in Pisa, Ricci always asked him to extend his regards to Betti and Dini, his former professors, and to mutual colleagues, including Bianchi, before signing his name. On one occasion, he asked Volterra to thank Bianchi for having sent him a recent paper, rather than writing directly to Bianchi himself. Did Ricci feel that enlisting Volterra as an intermediary would influence Bianchi's opinion of his absolute calculus? Although a sprinkling of correspondence between Ricci and his colleagues exists in various archives, historians have searched in vain for evidence of any letters exchanged between Ricci and Bianchi. In the absence of such documents, perhaps Ricci did rely on colleagues, including Volterra, to transmit his messages, such as this one, contained in a letter written in the summer of 1888. After asking Volterra to send his greetings to Bianchi, he wrote, "and tell him that I would really appreciate it if he found the time to take a look at my last memoir on covariant and contravariant ["Delle derivazione covarianti e controvarianti e del loro uso nella Analisi applicata"] differentiation, because I am hoping that he can find some advantage in the methods presented there for his studies in differential geometry." There is no indication that Bianchi ever replied.[38]

Bianchi's official opinion of Ricci's oeuvre surfaced more than a decade later, in connection with the latest Royal Prize in Mathematics, the last competition Ricci would ever enter. As the December 1901 deadline for entering the contest approached, eight applicants presented their publications for consideration, including Levi-Civita's friend from Bologna, Federigo Enriques, and Enriques's colleague and future brother-in-law, Guido Castelnuovo. The oft-told story of their collaboration had begun in Rome, in 1892, when Enriques, a gifted geometer who had earned his degree in mathematics at the Scuola Normale in Pisa, came to Rome to attend a course in higher geometry taught by Luigi Cremona, the founder of the Italian school of geometry. However Enriques found the much younger Castelnuovo, only five and half years his senior and already holding the university's newly created chair of analytical and projective geometry, a much more congenial mentor than Cremona, whose lectures completely befuddled him. The two young mathematicians struck up a firm friendship, cemented by evenings spent strolling the streets of Rome, engaged in rapt discussions of algebraic geometry. Many years later, in a eulogy for Enriques delivered at the Lincei after his death in 1946, Castelnuovo remarked, "It is probably not an exaggeration to assert that the theory of algebraic surfaces from the Italian point of view was created during these conversations."[39]

By 1900, their collaboration had reached an important milestone in their classification of algebraic surfaces, one of the great achievements of algebraic geometry. A year later, when they decided to try for the Royal Prize, they did so with the express stipulation that the fifteen papers they were submitting be judged as a unitary entry, even though only three had been coauthored. The other twelve, six by Enriques, six by Castelnuovo, were to be treated as part of their common entry, they said; and they would not entertain the notion of being judged separately. This unorthodox proviso set off alarms at the academy.

Before then, there had been joint winners, including Bianchi and Pincherle a decade earlier and Volterra and Corrado Segre in 1895, but these mathematicians had entered separately, and the prize had been divided between them. As always, the prize commission moved rather slowly; and it was only a year later, before deliberations had even begun on all the works under consideration, that Pietro Blaserna, the vice-president of the Lincei, took the unusual step of asking yet another administrative body, the council of state, for an opinion as to the admissibility of Enriques and Castelnuovo's unprecedented request. According to Italian mathematics historian Umberto Bottazzini, the council's response, issued in January 1903, "seemed to be favorable to Enriques and Castelnuovo, stating that in principle a joint application was admissible."[40] Some of the Royal Prize commissioners, however, dissented—most notably, Ricci's sometime scourge, sometime supporter Giuseppe Veronese, who had lost no time in requesting a copy of Castelnuovo and Enriques's dossier so that he could make a thorough investigation of it.

By the time the commissioners met again in April, a number of them seemed disinclined to challenge the "pressure of Veronese," who had suggested that they not only reject Enriques and Castelnuovo's conjoined submission outright but also take them up on their ultimatum and boot them from the competition altogether. "[We] must not," he exhorted his fellow commissioners, "establish a serious precedent, which allows researchers to join together in different fields, with works not prepared in common,"[41] and he further urged that deliberations be tabled for a year. Valentino Cerruti, another member of the commission and a mathematician at Rome, also worried that granting the request of Castelnuovo and Enriques would set a bad example. In theory, he pointed out, "Ricci and Levi-Civita could [also] present works done together at the end [clearly a reference to the recent *Annalen* article] and combine all the previous works prepared separately."[42] A zealous defender of the Royal Prize, perhaps because he had previously failed to win it himself, Veronese then prepared a comprehensive report on the Castelnuovo-Enriques situation, concluding with the commission's recommendation to postpone the competition. This was duly presented and approved at a general meeting of the Lincei Academy on June 6, over the objections of both Volterra and Segre.

Meanwhile Bianchi, the designated reader for the papers of Ricci and Ernesto Pascal, one of the other applicants, had already started examining their writings. Soon after, Enriques wrote to Castelnuovo sharing some gossip imparted to him by Levi-Civita, who had picked it up while visiting colleagues in Pisa: "The worst [news] is that Bianchi (and already before Veronese) pointed out [to Levi-Civita] the difficulty of our position as joint applicants. . . . The best [news] is that Bianchi himself let slip the secret that the scientific question [will] be between us and Ricci; they regard Pascal, so it seems, as virtually eliminated."[43]

Enriques's assessment of Ricci's chances may have been premature, judging from the final text of Bianchi's report. It unfolded methodically, beginning with one of the five works Ricci had submitted, his volume of lithographs on the geometry of surfaces. Bianchi described the *Lezioni* as "the fundamental work,"[44] offering a "clear and sober presentation" of the concepts and algorithms of "the new calculus" in its first 140 pages. Christoffel had introduced the covariant derivation, and Ricci, added Bianchi, deserved the "principal credit" for recognizing "the importance of this fundamental concept and building his new algorithms upon it." So far, so good. However, the balance of Ricci's book, dealing with problems of surfaces as flexible and inextensible to special classes of surfaces of the second degree, presented Bianchi with a clear conflict of interest, a fact he seemed inclined to ignore. Much of Bianchi's early work had been on the properties of surfaces, a subject in which he never lost interest; and his textbook *Lezioni di geometria differenziale*, based on his classroom lectures, covered much of the same ground as Ricci's treatise. Ricci's volume may have received a warm reception in America, but Italy's premier differential geometer wasn't about to let Ricci's tools loose in his own backyard. As he opined, "It would be fruitless to report in detail on the various chapters from [Ricci's] course, which, save for the different arrangement and in the particularity of the algorithms used, do not differ substantially from other textbooks." Here and there he found some "noteworthy additions" to the literature in Ricci's writings while maintaining that these particular results did not seem "to have the importance attributed to them by the author."[45]

Overall, he praised Ricci's book for the systematic presentation of the absolute differential calculus, but questioned the "excessive preponderance given to the algorithmic part," which often left "the essential geometric content in the shadow" and led Ricci on occasion to view geometrically self-evident properties as "special results of his calculus." In the same vein, Bianchi argued that analyzing the book from a geometric perspective exposed various deficiencies. Could these shortcomings, he asked rhetorically, be overcome by the "greater power and versatility" demonstrated by the new methods as compared with the old ones? His answer to his own question was predictable: "In the field of the infinitesimal geometry of surfaces, the inherent advantages of the new algorithms are mostly formal or limited to secondary questions," he wrote, adding that Ricci's new methods had contributed nothing significant to recent and important progress in the field. He conceded that in the field of multi-dimensional geometry, "where ordinary intuition becomes more difficult.... the procedures of the absolute calculus," frequently led not only to "simplified and condensed" calculations, but certainly more "direct" ones. The proof, he added, could be found in a paper of "considerable originality and generality of results," published by Ricci in 1896,[46] one of two articles on this topic he submitted for the prize competition.

Bianchi made short shrift of Ricci's last entry, a brief manuscript on the general theory of elasticity, by remarking that the results, all known and derived courtesy of Ricci's absolute calculus, "did not offer in substance any novelty," nor did they seem to have any "noticeable advantages" over the "usual one," which for "simplicity of ideas and clarity of formulas"[47] covered all the bases.

His analysis of Ricci's works completed, Bianchi turned to the task of summarizing it all. After stating for the record that only "a truly exceptional work"[48] merited the prize (which would finally be awarded in 1905 to Castelnuovo and Cesare Arzelà, professor of mathematics at Bologna, and in 1907, to Enriques and Levi-Civita), he concluded that neither Pascal's works, nor Ricci's for that matter, had reached the requisite lofty heights. It was the same old story: Ricci's work, while commendable and even sporadically original, was not of the overall caliber Italy expected of her finest mathematicians. Ricci's algorithms, Bianchi kindly noted, "developed and perfected by him with constant study served a useful [function]... in the treatment of various mathematical questions." However, they were not, in Bianchi's estimation, "indispensable." It was his considered judgment that while "the truly new results" in the work could not currently be deemed prizeworthy, such recognition "could be awarded" at some future time, provided they should be "suitably completed and perfected. . . ."

CHAPTER 10

# The Indispensable Mathematical Tool

"Grossmann, you must help me or else I'll go crazy."

A. Einstein to M. Grossmann, summer 1912[1]

In 1912, Albert Einstein, then ensconced at the Swiss Federal Institute of Technology (ETH) in Zurich, hit a wall in his efforts to incorporate gravity into his special theory of relativity. He had begun to lay the foundation for general relativity in 1907 with his principle of equivalence and slowly over the next several years, by restricting himself to a constant acceleration field, succeeded in formulating a static and relativistic theory of the gravitational field. The theory describes the effects of an unchanging, homogenous gravitational field, which is the case to a good approximation of the effect of gravity within a laboratory on Earth. "These are solid pieces of theoretical analysis,"[2] writes Abraham Pais referring to these early papers of Einstein, although, he adds, "it takes some time to grasp their logic."

And while, despite many efforts, a dynamic theory of gravity remained outside Einstein's grasp, he had discovered the year before "in the quiet rooms of the Institute of Theoretical Physics at the German University of Prague," that the principle of equivalence "demand[ed] a deflection of the light rays passing by the sun with observable magnitude"[3]—a prediction of 0.83 seconds of arc[4] in the vicinity of the sun's gravitational field—which he believed would provide a "starting-point for the theoretical understanding of gravitation."[5] However, astronomers on both sides of the Atlantic told him they needed a full solar eclipse to observe the almost imperceptible bending of light from the nearest stars, and the next one was not due until 1919.[6] Einstein resigned himself to waiting another eight years for the right viewing opportunity to present itself.

Four years earlier and closer to home, Hermann Minkowski, an extraordinarily gifted mathematician whose talents extended to number theory, geometry, and mathematical physics, had put mathematical teeth, so to speak, into Einstein's special relativity. After teaching for several years at the Zurich Polytechnic, where Einstein had been one of his students, Minkowski became a professor of mathematics at Göttingen, where he went on to develop and publish two major books on the geometry of numbers. Although Einstein had registered for eight courses with Minkowski, he often failed to attend classes,[7] and the rapport between professor and pupil apparently left something to be desired. While recognizing the importance of special relativity, Minkowski seems to have put little stock in his former student's mathematical abilities. "For me [special relativity] came as a tremendous surprise," he once confessed to physicist Max Born. "Einstein had been a lazy dog. He never bothered about mathematics at all."[8]

Minkowski did. His own ideas on relativity and spacetime took root in the advanced seminar on the equations of electrodynamics that he and Göttingen mathematical luminary David Hilbert taught in 1907 and where Einstein's 1905 relativity paper was one of the topics discussed. In formulating special relativity, Einstein had based his theory on two assumptions. The first is that the speed of light is a universal constant, independent of the speed of the observer. This had a number of startling consequences, including the fact that the passage of time itself becomes dependent on the speed of the observer. It ultimately led to Einstein's most famous equation, $E = mc^2$, which tells us that when a given amount of mass is totally converted to energy, that energy will be equal to the mass multiplied by the speed of light squared. The second assumption is that if two observers are traveling at a constant velocity, all they can establish is their relative motion.

A mathematician of incisive geometric insight, in 1908 Minkowski proceeded to reformulate Einstein's theory as a geometrical theory of four-dimensional spacetime. In his opening remarks on special relativity delivered before the Göttingen mathematical society in November 1907, he argued that the mathematician is exceptionally "well prepared to pick up the new views of space and time because it involves acclimating himself to conceptual schemes with which he has long been familiar." Physicists, he maintained, are compelled to discover these concepts afresh and must painfully cut their way through a jungle of obscurities. Meanwhile the familiar, well-trodden, pathways of the mathematician comfortably lead forward.[9]

Although few details of his seminar talk have survived, it's possible that Minkowski's remarks caused his listeners to sit taller in their chairs, as they took in his assurances that they had a formidable role to play in putting special relativity on a sound mathematical footing. If his geometrical approach "correctly reflects the phenomena," Minkowski said, it would demonstrate "the greatest triumph applied mathematics has ever shown." He then came to the heart of his argument: "Expressed as briefly as possible, it is this—the world in space and time in a certain sense is a four-dimensional, non-Euclidean manifold." Not only are the laws of physics most completely understood in four-dimensional spacetime, he told his audience, but physical reality itself is a mathematical structure.

In the introduction to his most famous talk, "Space and Time," delivered in 1908 at the eightieth meeting of the Assembly of Natural Scientists and Physicians in Cologne, Minkowski stressed the revolutionary character of his views: "Gentlemen! The views on space and time that I wish to lay before you have sprung from the soil of experimental physics. Therein lies their strength. They are radical. Henceforth space by itself, and time by itself, are doomed to fade away into mere shadows, and only a union of the two will preserve an independent reality."[10] In science as in politics, a revolutionary idea often demands its own vocabulary, which Minkowski willingly supplied as he described his investigation "along a purely mathematical line of thought to arrive at changed ideas of space and time." Such terms as world point, world line, absolute world, backward (and forward) lightcone, time-like vectors, and space-like vectors populate the text (the lecture was published in 1909), along with his famous spacetime diagram, which appeared for the first time as a lecture slide in this talk.

Although it would later be regarded as a landmark of relativistic literature, Minkowski's talk had a mixed reception. Mathematicians tended to respond positively, particularly those in Göttingen, who enthusiastically talked it up. Physicists were not so easily persuaded. According to historian Scott Walter, many took their time adjusting to the notion "that Euclidean space was no longer adequate for understanding physical phenomena."[11] Turning to Minkowski's legacy, Walter has this to say: "His lecture created a scandal for physicists in its day, but unlike most scandals, it did not fade away with the next provocation. Instead, Minkowski focused attention on how mathematics structures our understanding of the physical universe, in ways no other writer had done since Riemann, or has managed to do since, paving the way for acceptance of even more visually unintuitive theories to come in the early twentieth century, including general relativity."[12]

Minkowski's "introduction of four-dimensional objects and concepts,"[13] historian Jean Eisenstaedt has written, left an indelible mark on the evolution of special relativity. He points out that Minkowski "replaced the notion of a point by that of a universe point... and that of a curve by that of a universe line.... Those universe points would later be called events, but the concepts would remain. It was simply a matter of conforming to perception and no longer speaking of space and time separately but rather, together: we only observe events. A universe point, an event, is a place-at-a-given moment, while a universe line is a set of universe points, a trajectory in spacetime." But, Eisenstaedt emphasizes, "Minkowski introduced relativity's fundamental tool: *proper* time, which he constructed from the line element or *metric* of space-time." Einstein, in Minkowski's opinion, "had not pushed his analysis to its ultimate consequences."

For a variety of reasons, Einstein was slow to accept Minkowski's spacetime geometry. In an early biography of the physicist, Rudolf Kayser (the son-in-law of Einstein's second wife, Elsa, he wrote under the pseudonym Anton Reiser) traced Einstein's early dismissive attitude toward "mathematical speculation" to his fascination with "the visible processes of physics."[14] Mathematics simply encompassed too many specialized fields of study, or so it seemed to the youthful Einstein. "I saw myself in the position of Buridan's ass, which was unable to decide upon any particular bundle of hay,"[15] he wrote in autobiographical notes prepared when he was in his mid-sixties.

Recalling his "excellent teachers" (among whom he singled out German mathematician Adolf Hurwitz as well as Minkowski), he reflected, "I should have been able to obtain a mathematical training in depth." However, the choice came down to working diligently on his mathematics courses or spending time in the physics laboratory, and he opted for the latter, "fascinated by the direct contact with experience."

By his own account, Einstein largely spent the balance of his free time reading physics textbooks at home. Physics, too, branched out in many directions, but his deep "interest in the study of nature" prompted him "to scent out that which might lead to fundamentals" and to discard "everything else, from the multitude of things that clutter up the mind and divert it from the essentials." But as a letter he wrote in 1912 to his Munich physics colleague Arnold Sommerfeld, suggests, Einstein's struggle to include gravity in the theory of special relativity would ultimately lead him to appreciate that better knowledge of basic physics depends on advanced mathematical methods.

By inclination Einstein adopted a cautious attitude toward complicated mathematical methods—he had relied on algebraic equations and simple calculus in his presentation of special relativity. His distrust of abstract mathematics may help to explain why he initially recoiled from adopting Minkowski's spacetime theory. His attitude toward Minkowski's work recalls Beltrami's semi-dismissive view of Ricci's absolute calculus at the turn of the century—that here was a highly complex, if not byzantine, analytical construct begging for simplification. For his part, aside from asking Einstein for a reprint of his first relativity paper—which he needed to prepare his 1907 talk to the Göttingen mathematical society—Minkowski does not seem to have ever corresponded again with his "lazy dog" of a student. Nor did they confer in person about their mutual interest in relativity theory. Minkowski was to die suddenly of appendicitis early in 1909, several years before Einstein's labors culminated in the theory of general relativity.

In the end, Einstein's reluctant capitulation to complex mathematics came from a different source. In early 1908, Jakob Laub, a young Austro-Hungarian physicist at the University of Wurzburg and a former student of Minkowski's at Göttingen, wrote to Einstein, then a sporadically compensated lecturer or *privatdozent,* at the University of Bern, asking if they could spend some time working together. An experimentalist with a keen interest in "relativity physics,"[16] Laub began a three-week collaboration with Einstein that April, during which they wrote two papers in response to Minkowski's paper on the fundamental equations for electromagnetic phenomena in moving bodies, published in 1908. In his discussion of Minkowski's detailed 1908 response to special relativity, Einstein biographer Abraham Pais notes that "for the first time the Maxwell–Lorentz equations are presented in their modern tensor form, the equations of point mechanics are given a similar treatment, and the inadequacy of the Newtonian gravitation theory from the relativistic point of view is discussed."[17] Einstein and Laub, however, took exception to Minkowski's sophisticated four-dimensional formalism in treating these topics. "In view of the fact,"[18] they wrote in the introduction to their first paper, "that this study makes rather great demands on the reader in its mathematical aspect, we do not consider it superfluous to derive here these important equations in an elementary way, which is, by the way, essentially in agreement with that of Minkowski."

"Laub is quite a nice man," Einstein confided in a letter to Mileva Einstein-Marić, his then-wife, before adding, "though very ambitious, almost rapacious."[19] In Laub, Einstein had found someone willing to do the calculations "which I wouldn't find time to do, and this is good." Laub, who like Einstein was initially not impressed with Minkowski's four-dimensional derivation of the Maxwell-Lorentz field equations, later sent word to Einstein that Moritz Cantor, a colleague at Wurzburg, disliked both their paper and Minkowski's. "When I asked him what in fact it means physically to treat time as a fourth spatial coordinate... he left my question unanswered,"[20] Laub reported. "I think he has been enormously impressed by non-Euclidean geometry." Later that summer, in a letter bemoaning the prospect of grappling with complicated mathematical derivations on his own, Einstein advised Laub that "all this horrible calculating will not make one any wiser."[21] Nevertheless within three years Einstein had adopted Minkowski's geometric four-dimensional spacetime formulation of special relativity.

In 1911, after several years of thinking about light quanta and other physical problems, Einstein had returned to an insight he later remembered as "the happiest thought of my life."[22] He had been "sitting on a chair in the patent office in Bern," he later told a group of students and faculty at the University of Kyoto. "Suddenly a thought struck me. If a man falls freely, he would not feel his own weight. I was taken aback. This simple thought experiment made a deep impression on me. This led me to the theory of gravity."[23] This is Einstein's equivalence principle: that a uniformly accelerated frame is exactly physically equivalent to an inertial frame subject to a uniform gravitational field. That realization paves the way for a description of the effects of gravity on a test mass as free fall in curved geometry. It also brings to the fore the problem of finding coordinate transformations between inertial and accelerated frames that preserve the form of physical equations, which are then designated as covariant.

Einstein carried that critical insight with him to his appointment that year as a professor at the German University of Prague. While there he became friendly with Georg Pick, a highly respected Austrian mathematician, formerly an assistant to the physicist Ernst Mach and now a senior member of Prague's science faculty. Like Einstein, Pick played the violin; he introduced his younger colleague to chamber music groups and they got together often to go on walks. As they discussed their research during these strolls, Pick's companion described his unsuccessful efforts to generalize his theory of relativity.

"Already at that time," according to Philipp Frank, one of Einstein's early biographers and his professorial replacement when Einstein left Prague, "Pick made the suggestion that the appropriate mathematical instrument for the further development of Einstein's idea was the 'absolute differential calculus' of the Italian mathematicians Ricci and Levi-Civita."[24] For whatever reason, perhaps because he did not yet perceive the lack of the right geometry to be his main stumbling block, Einstein did not rush off to the library. Nor is it likely that Pick, twenty years Einstein's senior and a stickler for protocol, volunteered to retrieve the work himself or to help the young physicist decipher its complex mathematical content. That July, in the last paper he wrote before leaving Prague, Einstein hinted in so many words that "to create a relativity-theoretical scheme in which the equivalence of inertial and gravitational mass finds expression"[25] would certainly require the use of new theoretical tools. Later in the same paper, he went on to discuss how

> the equivalence principle opens up for us the interesting perspective according to which the equations of a relativity theory that would also include gravitation may also be invariant with respect to acceleration (and rotation) transformations. In any case, the road to this goal seems to be a quite difficult one. One can already see from the previously treated, highly specialized case of the gravitation of rest masses that the space-time coordinates will lose their simple physical meaning, and it is not yet possible to foretell the form that the general space-time transformation equations can have. I would like to ask all of my colleagues to have a try at this important problem!

No one would try harder than Einstein himself over the course of the next three years.

In 1912, Einstein left Prague and settled into a new job at the Swiss Federal Institute of Technology, ETH, in Zurich. There had been some opposition to his joining the faculty there, but once again he proved fortunate in his friends. The invitation to join ETH had come from a former classmate, Marcel Grossmann, a mathematician of outstanding ability and uncommon generosity, who had already played a pivotal role in shaping Einstein's scientific career and was destined to do so again. As the recently installed head of the ETH's physics and mathematics department, he was in a position to see the appointment through, and in August of that year Einstein joined ETH as its first faculty member in theoretical physics.

For Einstein the return to Zurich and the resumption of his association with Grossmann was a homecoming of sorts. The two had first met in the city fifteen years earlier as students at ETH's forerunner, the Swiss Polytechnic School, where they bonded over a shared love of physics and mathematics, as well as a fondness for amateur psychology. Outside the classroom, the two young men spent hours sipping iced coffee at the city's Café Metropole, holding forth about "everything that could be of interest to young folks with open eyes,"[26] and, as Grossmann recalled some years later, "psychologically analyzing joint acquaintances as well as ourselves."[27]

Their backgrounds were similar: both came from solidly bourgeois, German-speaking Jewish families that placed a high value on cultural and intellectual attainments. Einstein hailed from Ulm in Southern Germany, where his father owned an electrochemical factory; Grossmann, who was born and spent his early years in Hungary, had ancestral roots in Switzerland and moved back to Basel with his parents as a teenager when his father secured a job managing a factory for agricultural machinery. Both showed precocious abilities as youngsters, but the very different routes they followed into the Polytechnic reflected some notable differences in experience and temperament. We do not know as much about Grossmann's youth[28] as we do about Einstein's, but he seems to have found it easy to adapt to the demands of his middle school in Basel, where he excelled as a student and, in a manner reminiscent of Levi-Civita's academic performance some years earlier in Italy, sailed through his entrance exams for the Polytechnic in 1895.

Einstein's educational trajectory was more turbulent. When he applied to the Zurich Polytechnic for the first time, he lacked a high-school diploma, having fled the Luitpold Gymnasium in Munich six months earlier to join his family in northern Italy, where his father and uncle had relocated their factory. His six years at the school had taught him to despise the pervasive militaristic atmosphere, with its emphasis on regimented and rote learning, and he found the prospect of fulfilling his own army obligations in a few years' time equally unsavory. On one occasion, Einstein's Greek teacher had assured him that he would never amount to anything and told him he should leave the gymnasium. When Einstein protested that he had done nothing wrong, the teacher retorted, "Your mere presence spoils the respect of the class for me."[29] Soon after, the "depressed and nervous" young Albert, according to his sister, Maja,[30] took it upon himself, with the help of a doctor's note, to withdraw from school, vowing never to return to Munich. (Later on, he also renounced his German citizenship, rendering him stateless for five years, until he had saved up enough money to become a Swiss citizen in 1901.)

On his first try to enter the Polytechnic in 1895, Einstein failed the entrance exam. He scored poorly on a range of general knowledge topics, ranging from literary and political history to botany and zoology, but did well enough in the scientific subjects—it helped that he had studied differential calculus and some physics on his own since the age of twelve—that one of the school's physics professors, Heinrich Weber, invited him to audit his second-year physics course if he remained in Zurich. Instead, Einstein opted to complete his secondary school education at Switzerland's first non-parochial high school, the technical branch of the Aargau Cantonal School, in Aarau, a small town west of Zurich. In September 1896, he sat for the school's comprehensive exams for the *matura* certificate—*the* ticket for automatic admission to the Polytechnic—followed by a series of ten-minute oral examinations open to the public. Although Hermann Einstein had become accustomed to wild swings in his son's report cards ("with Albert I got used a long time ago to finding not-so-good grades along with very good ones,"[31] he once remarked), Einstein came in first on the written examinations, despite having managed to arrive late for the physics test, then departing early.

Although Einstein found the social and intellectual atmosphere at Zurich Polytechnic more congenial than that of Luitpold Gymnasium (ironically, to later be renamed the Albert Einstein Gymnasium in honor of its renowned non-graduate), his academic performance probably came, once again, as no surprise to his father. Within a year of his arrival, he had alienated Weber, the physics professor who had earlier offered him support, and his indifference to subjects that did not personally interest him (of which there were many) led him to skip a sizable number of his classes, which, as Minkowski's characterization of him as an indolent hound suggests, did not endear him to the faculty. Grossmann, by contrast, had quickly established himself as a dedicated and exceptional student. On more than one occasion, Einstein borrowed his friend's exquisitely recorded mathematics lecture notes, taken in class and meticulously written out in a clear and legible hand,[32] to prepare for his examinations.

Grossmann also brought Einstein home to meet his family, who upon hearing from Marcel's own lips, "this Einstein will one day be a very great man,"[33] encouraged him to visit often. When he learned from his son that Einstein's spotty record at Swiss Polytechnic had made it highly unlikely that he would ever find a postgraduate position as a university assistant, the elder Grossmann recommended him to a friend who was the director of the Swiss federal office in Bern. Einstein's appointment as technical expert third class at the patent office, initially on a trial basis, was made in June 1902; it became permanent two years later. It was there that Einstein completed the theoretical work that in 1905 culminated in a quartet of papers on the photoelectric effect, Brownian motion, special relativity, and mass-energy equivalence, published during what became known as his *annus mirabilis* or "miracle year," and launching a revolution in physics that continues to this day.

When Grossmann died from complications of multiple sclerosis in 1936, Einstein wrote a touching letter to his widow. Of his student days, he remarked: "He, the irreproachable student, I myself, disorderly and a dreamer. He, on good terms with the teachers and understanding everything, I a pariah, malcontent and little loved. But we were good friends and our conversations. . .in the Metropole every few weeks are among my happiest memories. Then, the end of our studies—I was suddenly abandoned by everyone, standing at a loss on the threshold of life. But he stood by me and thanks to him and his father [Julius Grossmann] I obtained a post later with Haller in the Patent Office. It was a kind of salvation and without it, although I probably should not have died, I should have been intellectually damaged."[34]

Having served as a willing if unwitting catalyst for Einstein's 1905 achievements, Grossmann went on to a more conventionally successful career, rising swiftly up the academic ladder. He wrote his Polytechnic diploma thesis on non-Euclidean geometry with Otto Wilhelm Fiedler, whose assistant and PhD student he then became, specializing in Fiedler's areas of descriptive and projective geometry. Later he taught at his former middle school in Basel, while also holding a position as privatdozent at the University of Basel. He published two geometry textbooks there, and when Fiedler went on medical leave from Polytechnic in 1906, Grossmann took his place provisionally and a year later was appointed Fiedler's successor. In 1911, the same year the Polytechnic changed its name to ETH, he became head of its physics and mathematics program and recruited Einstein onto the faculty.

Now, reunited with Grossmann at ETH, Einstein described his current mathematical struggles to his old friend, explaining that although Euclidean geometry had worked for treating special relativity, he would have to go beyond it to describe motion in the presence of gravitational forces. For a theory in which masses cause paths to curve, he needed a mathematical tool that would "furnish a complete solution to cases where the force of gravity had different directions at different points in space."[35] Finally he blurted out, "Grossmann, you must help me or else I'll go crazy!"[36]

To his eternal credit, Grossmann did not remind Einstein that during their undergraduate days, Einstein, in his capacity as a *kaffeeklatsch* analyst, had loftily advised his friend that "your main weakness is, you cannot say 'no.' "[37] Rather, as Einstein recounted decades later, Grossmann responded with alacrity to Einstein's request that he "please go to the library and see if there exist[s] an appropriate geometry to handle such questions." "The next day," as Einstein told his future biographer, Abraham Pais, "Grossmann returned and said that there indeed was such a geometry, Riemannian geometry."[38]

Ernst Straus, a mathematician who had worked with Einstein at the Institute for Advanced Study in Princeton ten years earlier, fleshed out Pais's recollections. According to Straus, "Grossmann not only told Einstein that Riemannian geometry was the answer to his problem," but cautioned him, "that [geometry] is a terrible mess which physicists should not be involved with."[39] Dismayed by that news, Einstein asked whether any other geometries might prove useful, to which Grossmann responded by shaking his head. He also "pointed out to Einstein that the differential equations of Riemannian geometry are nonlinear, which he considered a bad feature." Not so, replied Einstein, who had already figured out that generalizing special relativity "would probably involve non-linear differential equations,"[40] the Einstein scholar Tilman Sauer has concluded. Grossmann, Einstein later recalled, "at once caught fire, although as a mathematician he had a somewhat skeptical stance towards physics."[41]

In the forward to the Czech translation of his first popular account of the special and general theories of relativity, Einstein recalled how it was only upon his return to Zurich that "I hit upon the decisive idea about the analogy between the mathematical problem connected with my theory [of general relativity] and the theory of surfaces by Gauss—originally without knowledge of the research by Riemann, Ricci, and Levi-Civita."[42] While Einstein's knowledge of nineteenth-century mathematical literature may have been sketchy, it was not for lack of exposure. As an undergraduate at Zurich's Polytechnic, he had been exposed to a great many required courses in mathematics, including Swiss mathematician Karl Friedrich Geiser's lectures—when he bothered to attend them—on analytical geometry and determinants in his freshman year, and a pair of classes, again by Geiser, on the geometric theory of invariants and on infinitesimal geometry, which covered Gauss's theory of surfaces and Riemann analysis. Much later in life he remembered the lectures in the latter course as "true masterpieces of pedagogical art, which later helped me very much in wrestling with general relativity."[43] While Gauss had pioneered the geometry of surfaces in Euclidean space, it was Riemann who had "studied the foundations of geometry in an even more profound way,"[44] Einstein told an audience in Kyoto years later, in the course of relating how his collaboration with Grossmann had begun.

Listening to Grossmann in 1912, Einstein had experienced a sudden flashback to Geiser's lectures on the Gaussian theory of surfaces. But it was Grossmann's discussion of Riemannian spaces—Riemann's mathematical techniques for building and working in curved spaces of any dimension (including, crucially, four)—that held Einstein in thrall. "I learned for the first time about Ricci and later about Riemann. So I asked my friend whether my problem could be solved by Riemann's theory." Within days of his frustrated outburst to Grossmann and his more subdued request that his old friend scour the mathematical literature, he had his answer: The German mathematician Bernhard Riemann had indeed provided the geometry he needed. Moreover, Grossmann told him, two contemporary Italian mathematicians—Gregorio Ricci and his former pupil, now colleague, Tullio Levi-Civita—had together developed the necessary mathematical methods to pose the equations of physics in Riemann's curved spaces. It was the beginning of Einstein's acquaintance with the absolute differential calculus.

Today, we might imagine an historic meeting among the principals—Einstein, Grossmann, and their Italian counterparts. Across the Alps, close by, if not within shouting distance, Ricci was at his stately country house, revising his classroom lectures on the mathematical theory of elasticity,[45] and blessedly free at last of preoccupations over long-denied promotions. His ebullient younger colleague, whose future would include an intense correspondence with Einstein, was gearing up to teach special relativity at the University of Padua. Such an encounter, alas, never took place.

What we do know for certain is that in the summer of 1912, shortly after his thirty-third birthday, Einstein's longstanding indifference to mathematics changed abruptly as he came face-to-face with Ricci and Levi-Civita's seventy-six-page article, published in Klein's *Annalen* twelve years earlier. In common with others who had engaged with the more abstruse properties of Ricci's mathematics, Einstein found the learning process difficult. From Zurich, on October 29, 1912, he wrote to Sommerfeld:

> I occupy myself exclusively with the problem of gravitation and now believe I will overcome all difficulties with the help of a friendly mathematician [Marcel Grossmann] here. But this one thing is certain: that in all my life I have never before labored at all as hard, and that I have become imbued with a great respect for mathematics, the subtle parts of which, in my innocence, I had till now regarded as pure luxury. Compared with this problem, the original theory of relativity is child's play.[46]

For perhaps the first time in his career as a theoretical physicist, Einstein realized, as he later said, "that the foundations of geometry have physical significance."[47]

Less than a year after Einstein first encountered the absolute calculus, his scientific collaboration with Grossmann culminated in a joint paper,[48] Published in the spring of 1913, it was an almost fully formed theory of gravitation, with a structure very similar to what would eventually be general relativity. The paper—later characterized by Einstein as a "feverish scientific work on the formalism of the general theory of relativity"[49]— reveals a clear division of labor between the two friends: the physical part belongs to Einstein and carries his byline; the mathematical part, authored and signed by Grossmann, provides a cogent and comprehensive account and expansion (in its application if not in its formalism) of Riemannian geometry and the absolute differential calculus developed by Ricci and Levi-Civita at the turn of the century. In a penetrating and lucid account of Grossmann's role in the joint Einstein-Grossmann collaboration, Sauer argues that "we must assume that Grossmann actually helped clarify the very mathematical status of the objects that were entering the center stage of their theoretical efforts."[50] He adds that

> [Grossmann] introduced covariant, contravariant, and mixed tensors for spaces of arbitrary dimensions and of any rank. The use of the word 'tensor' in this context is a novelty. Ricci and Levi-Civita had called these objects *systèmes covariants ou contrevariants,* [covariant or contravariant systems], and they had never considered *systèmes* of mixed transformation behavior, i.e., with a mix of covariant and contravariant indices.... Grossmann introduced a notation where all indices were written as

> subscripts and the transformation character was indicated by writing the object itself with a Latin, Greek, or Gothic character for covariant, contravariant, or mixed tensors, respectively.

In the hands of Grossmann and Einstein, Ricci and Levi-Civita's absolute calculus became the indispensable tool to write covariant equations of physics, but for a variety of reasons Einstein initially concluded that it was impossible to construct *generally* covariant gravitational field equations (that is, equations that maintain their form with respect to *any* coordinate transformation) that would satisfy all the physical requirements of the theory of general relativity. When he changed his mind in November 1915,[51] he touched off a surge of interest among physicists and mathematicians alike, who rushed to the academic libraries seeking their own copies of Ricci and Levi-Civita's landmark paper detailing the absolute differential calculus, today more generally known by the name Grossmann first gave it and that Einstein made famous—the tensor calculus.

What was it about the tensor calculus that made it so useful for Einstein's purposes? On the physical side, Einstein had asked Grossmann for help in finding the particular mathematical equations that would account for the effects of gravity on matter and light. The geometry of special relativity is flat (like ordinary Euclidean geometry), even if thinking of space and time as a single entity yields counterintuitive consequences such as time dilation and the relativity of simultaneity. In contrast, the inclusion of gravity requires the curved spaces (and now spacetimes) described by Riemann. Riemannian spaces (or manifolds) are determined fully by specifying the Riemannian metric: a positive quadratic differential form that describes the distance between infinitesimally close points.

What Ricci and Levi-Civita brought to this heady mix of physical theory and mathematics was the notion of generally covariant "functions," now known as tensors. In the absolute calculus, as Levi-Civita wrote, "the formulas and results always remained in the same form, no matter what system of coordinates are being used. This explains why we have... a system of functions [tensors] that behave in the same way when the coordinates are changed, independent of the choice of these coordinates. In addition, certain operations are introduced that are equally independent of the coordinates chosen, i.e. they are absolute, giving the name to the calculus."[52] The most useful applications, Levi-Civita continues, "arise when the nature of the material under consideration requires a quadratic differential form... in general relativity, the infinitesimal interval between two events in spacetime. It is convenient then to take this differential form as given—that is, as absolute—and this is where the essential element of the new calculus arises: in the notion of covariant differentiation, which has the essential characteristics of ordinary differentiation, but also respects the invariant behavior (i.e., is independent of the choice of the coordinates) of the system to which it is applied." Differentiation (i.e., derivatives) is needed to describe changing positions, velocities, temperatures; in short, to do physics.

Einstein's audacious goal was to "unify all physical phenomena, connecting geometry and gravitation,"[53] Levi-Civita has remarked. How did he translate this goal into mathematical equations? Jean Eisenstaedt explains: "The fundamental equations of general relativity—the field equations—are a sort of machine to define the spacetime whose curvature represents gravitation."[54] He goes on: "In this complex mathematical mechanism, the right-hand side of the equation (also called the matter tensor) represents the distribution of matter, while the left-hand side represents the geometric structure of space by means of variables known as gravitational potentials."[55] The last step involves solving "the series of rather complex equations so obtained... and the solution is a curved spacetime." As one contemporary physicist has said, "The central idea of general relativity is that gravity arises from the curvature of spacetime. Gravity *is* geometry."[56]

In 1916, Einstein published the fundamental paper on the theory of general relativity—which made a number of predictions. In 1919, two astronomical expeditions independently confirmed one of them—the bending of starlight in the vicinity of the sun during an eclipse. This dramatic news—which made the front page of the New York Times almost single-handedly turned Einstein the theoretical physicist into a twentieth century icon. When one of his sons asked him around this time why he was so famous, Einstein, whose first glimmers of special relativity had come to him as a boy of sixteen when he imagined himself chasing, then riding, a beam of light, used a simple metaphor to illustrate his subsequent insight that matter and energy bend the geometry of the universe out of shape, producing the effect we experience as gravity. "When a blind beetle crawls over the surface of a curved branch, it doesn't notice that the track it has covered is indeed curved," he said. "I was lucky enough to notice what the beetle didn't notice."[57]

*Michele Valisneri provided considerable help with the technical sections of this chapter.*

CHAPTER 11

# "Write To Me Next Time In Italian"

"We are all in your debt,"[1] Levi-Civita wrote in early 1908 to the German physicist Max Abraham, a theoretician he was eager to have attend the Fourth International Congress of Mathematicians in Rome later that year. It was an assessment many of his colleagues would have echoed, but they probably would have added that it was easy for Levi-Civita to bestow such praise as he had never met the individual in question personally. Abraham was born in 1875 into a prosperous Prussian Jewish family in Danzig, Germany (now Gdansk, Poland). He attended the University of Berlin, where he studied under Max Planck, later to achieve fame for introducing quantum theory. Despite earning a PhD in 1897 in theoretical physics and serving for the next three years in the prestigious position of Planck's assistant at the university's Institute for Theoretical Physics, Abraham left Berlin hat in hand, an unemployed theorist. He spent the next nine years as a lowly *privatdozent* at the University of Göttingen while being repeatedly passed over for an academic position in any one of Germany's two dozen universities and technical institutes.

Abraham had a sharp scientific mind, matched, often to his detriment, by an equally sharp, biting tongue. Describing their colleague in later years, the physicists Max Born and Max von Laue recalled that "clarity was the essence of his being."[2] If colleagues indulged in silly or illogical arguments, Abraham would go on the attack, haranguing and mocking them publicly, "often in an exaggerated way." His penchant for commenting openly on his peers' personal foibles succeeded in alienating many of the very people who might have considered hiring him. As Born and von Laue recalled, "he didn't spare anyone"—an acknowledgment perhaps that they too had been on the receiving end of his trademark acerbic remarks. If Abraham thought of a joke at someone's expense, he told it on the spot—to do otherwise, his Göttingen colleagues remarked, "was beyond his control." However, none of this seemed to matter to Levi-Civita when he finally met Abraham at the mathematical congress. He took an instant shine to him.

Many months earlier, Guido Castelnuovo, the secretary of the congress's organizing committee, had enlisted Levi-Civita's help in selecting foreign speakers for the session devoted to theoretical mechanics and mathematical physics. Already conversant with Abraham's mastery of James Clerk Maxwell's electromagnetic equations and with his two-volume textbook on electrodynamics, *Theorie der Elektrizität,* which went through numerous editions[3] and was a staple of German universities' physics classes for many years, Levi-Civita wrote to him, extending an invitation to speak on his theory of the electron.

Largely forgotten today, except by historians of science, Abraham's model of the electron attracted considerable attention at the time. The particle's properties had become a subject of intense interest in Europe's physics capitals following its discovery by the English physicist J. J. Thomson barely ten years earlier. Abraham's theory posited that the mass of the electron was purely electromagnetic in origin. He characterized the particle as a completely rigid sphere whose negative charge was equally distributed throughout its surface. The calculations that he performed in Göttingen to arrive at this structure seemed to be in good agreement with the results of a series of experiments carried out earlier by his Göttingen colleague Walter Kaufmann on the electron's inertia and mass and, for a time, lent considerable weight to the theory.

Abraham replied in the affirmative to Levi-Civita's invitation but with a caveat.[4] He would talk about his electron model but judged it premature to provide a precise title. The field itself was moving very fast, and theories about the structure and nature of the electron were changing rapidly. In 1904, the renowned Dutch theorist Hendrik Lorentz had formulated a competing model in an effort to explain away the result of the famous Michelson–Morley experiment on ether drift. The experiment's findings had posed a direct threat to Maxwell's theory of the propagation of light, which postulated that light waves travel through an unseen "ether" that permeates all of space. Lorentz hypothesized that the dimensions of the electron and other solid bodies are slightly contracted by their motion through the ether. This theory was also wrong, but in his efforts to relate measurements of distance and time, he derived a set of equations that relate the measurements of two observers. These equations, known as the Lorentz transformation, would later figure prominently in Einstein's special theory of relativity. However, Michelson never abandoned his beloved ether, and neither did Lorentz or Abraham. Special relativity, by contrast, had no need of it, one reason for Abraham's immediate dislike of the theory.[5]

On April 5, 1908, more than five hundred mathematicians from twenty-two countries assembled in Rome for the International Congress. Many of the discipline's leading figures, including Lorentz, Ernst Mach, Gösta Mittag-Leffler, Émile Picard, Max Noether, and his daughter, Emmy (whose renown would one day rival her father's), gathered at a reception in the great hall of the University of Rome that evening (although it was later rumored that some of the delegates had skipped the lectures in favor of touring the Forum, the Coliseum, and other fabled attractions of the ancient city). Italy's King Victor Emmanuel III attended the opening session, which featured the Italian mathematician and statesman Vito Volterra, both the indisputable head of the Italian school of mathematics and a senator in Italy's parliament.

In his address, Volterra traced the progress of mathematics in Italy from the peninsula's unification to the opening years of the twentieth century, acknowledging the work of dozens of his contemporaries, most of whom he knew personally. He cited Levi-Civita's work on analytic mechanics and elasticity and commended Ricci for the introduction of "new methods"[6] that had complemented Bianchi's "many important and brilliant contributions" to almost all branches of differential geometry. Volterra's faintly patronizing assessment echoed the mathematical community's general opinion of Ricci's work at the time; Bianchi had said much the same thing about Ricci's absolute differential calculus in his official report on Italy's Royal Mathematics Prize in 1904. It is probably not what Ricci, who attended the meeting with his wife, Bianca, thought about his algorithms. His reaction to Volterra's address is unrecorded: validation and recognition still lay a few years in the future. His Paduan colleague and one-time academic rival, Giuseppe Veronese, enjoyed a much higher profile in Rome, delivering a plenary talk on non-archimedean geometry on the last day of the conference. Abraham's brief and somewhat esoteric talk, on which he had bestowed the title "On the Theory of Vortex Brakes," sank like a stone under a sea of unread conference reports.

A few months later, still unable to find a job in Germany, even after the Prussian government conferred on him the honorary title of professor, Abraham sailed away to America, where he had accepted a two-year professorial appointment at the University of Illinois in Urbana–Champaign. However, he returned to Germany after only one semester, disenchanted with the parochial atmosphere of the town and its university, his heavy teaching load, and the preponderance of poorly prepared students.[7] The famously sarcastic professor had no advocate to advance his interests in Europe, so Levi-Civita appointed himself to the position. In 1909, he wrote to a colleague at the University of Turin, possibly the mathematical physicist Carlo Somigliana, and urged him to consider Abraham for a vacant position there. In his letter, Levi-Civita also passed on a remark made by the German mathematician Adolf Kneser to the effect that Abraham's personality was not the only problem his "witty and agreeable" colleague faced in Kaiser Wilhelm's Germany. According to Kneser, "Abraham had made himself disliked by all the experimental physicists because he had repeatedly expressed himself in a manner that was rather unflattering to the 'big-shots' and that for this reason (in addition to widespread anti-Semitism, which we can consider quite probable there), it would have been very difficult for him to become a professor in a German university."[8]

The Turin appointment failed to materialize, but in 1909, possibly encouraged and certainly endorsed by Levi-Civita, who was on the hiring committee, Abraham entered and won the *concorso* for a position as associate professor of rational mechanics at the Milan Polytechnic Institute. "We find before us a scientist of notable fame,"[9] the committee noted, according to Levi-Civita, who took notes for the group and wrote the final report. The report continued, "His research on electromagnetic dynamics, in which a mastery of classical mechanics is singularly evident while also encompassing subsequent developments, serves as the solid and systematic theoretical foundation for the brilliant conceptions of modern physics." In an era when theoretical physics had yet to gain a foothold in the Italian academy, the mathematicians on Abraham's search committee saw his appointment as a means to bridge the gap created by the absence of such scientists in Italy, a vacuum that persisted down to the time of Enrico Fermi. Almost by default, Abraham became Italy's premier theoretical physicist.

Although an immensely talented scientist, Abraham was shortly to find himself "at scientific odds with Einstein, in regard both to the special theory and the general theory of relativity—and to lose in both instances,"[10] writes Abraham Pais. By the time he joined the faculty at Milan's Polytechnic in 1909, experimentalists in England and Germany had successfully challenged Kaufmann's analysis of his data on the mass of the electron, essentially shutting down Kaufmann's forays into this branch of modern physics. Even as others lost interest in Kaufmann's results, Abraham remained their advocate. As Born and von Laue recall in their sober but sympathetic obituary of Abraham, "he was an honorable opponent who fought with honest weapons and who did not cover up a defeat by lamentation and nonfactual arguments. The abstractions of Einstein were deeply repugnant to him; he loved his absolute ether, his field equations, his rigid electron, as a youth does his first flame, whose memories cannot be erased by later experiences."[11]

Even before leaving Germany for Italy, Abraham had become an outspoken critic of Einstein's special theory of relativity.[12] He continued to play this role in Milan, where he seems to have divided his time between teaching rational mechanics and taking potshots at Einstein's efforts to generalize the theory. Meanwhile, its creator, recently named a full professor at the German University in Prague, was lecturing to "less intelligent and industrious"[13] students than he had encountered in Switzerland and working furiously ("driven like a horse,"[14] he told a friend) to develop a comprehensible theory of the static gravitational field. Shortly after Einstein's first paper on the subject, "On the Influence of Gravitation on the Propagation of Light," appeared in the pages of the *Annalen der Physik,* he wrote to Laub, "The relativistic treatment of gravity is causing serious difficulties,"[15] before adding, "I consider it probable that in its customary formulation the principle of the constancy of the velocity of light holds only for spaces of constant gravitational potential."

Early in 1912, about six months after Einstein's paper appeared, Abraham responded with a theory of gravitation of his own,[16] which Einstein "at the first moment (for fourteen days!)"[17] found convincing. For those two weeks, he later recalled, "I too was totally 'bluffed' by the beauty and simplicity of his formulas." Within another fortnight, however, Einstein's ardor had cooled. "Abraham has supplemented my gravitation thing... but he made some serious mistakes in reasoning so that the thing is probably wrong,"[18] he wrote to one correspondent, adding, "This is what happens when one operates formally, without thinking physically!" In February, he used stronger language in a letter to the physicist Paul Ehrenfest, calling Abraham's theory "completely untenable."[19] That same month, Einstein took the rival theories public, noting in his second *Annalen* paper on gravitation that Abraham's arguments contradicted the "equivalence hypothesis and that his conception of time and space does not hold up even from a purely formal, mathematical point of view."[20] In their subsequent volley of *Annalen* notes, both physicists conceded minor points, with Abraham taking increasing aim at the theory of relativity itself. In one note, as Carlo Cattani and Michelangelo De Maria point out in their discussion of the Einstein-Abraham contretemps, "[Abraham] observed quite rightly that Einstein had not applied [the principle of equivalence] thoroughly and with coherence to his own theory and that therefore Einstein's theory was founded upon 'shaky ground.'"[21] Endless fascination with the theory of relativity, he warned the journal's readers, might well pose a threat to "the healthy development of theoretical physics."

In his reply, Einstein took issue, firmly but rather gently, with Abraham's claim that he [Einstein] had "delivered the coup de grâce to the relativity theory by abandoning"[22] the principle of constant $c$ [i.e., the speed of light]. "The theory is correct to the extent to which the two principles upon which [the theory] is based are correct.... Since these seem to be correct to a great extent, the theory of relativity in its present form seems to represent an important advance; I do not think that it has hampered the further development of theoretical physics." He also emphasized that relativity theory in its original formulation (special relativity) would always remain valid as the simplest theory in the case of a constant gravitational potential. As to what lay ahead, he remarked,

> It must be a task of the immediate future to create a relativity-theoretical scheme in which the equivalence of inertial and gravitational mass finds expression. I sought to make a first, quite modest contribution to the attainment of this goal in my papers on the static gravitational field. There I started...by conceiving the static gravitational field as physically identical with an acceleration of the reference system. I have to admit that I was able to carry through this conception in a consistent way only for infinitely small spaces, and that I cannot give any satisfactory reason for that fact. But I do not see this as any reason to reject the equivalence principle for the infinitely small as well; no one can deny that this principle is a natural extrapolation of one of the most general empirical laws of physics. On the other hand, the equivalence principle opens up for us the interesting perspective according to

> which the equations of a relativity theory that would also include gravitation may also be invariant with respect to acceleration (and rotation) transformations. In any case, the road to this goal seems to be a quite difficult one. I would like to ask all of my colleagues to have a try at this important problem![23]

If Einstein intended by this remark to extend an olive branch to Abraham, it had the opposite effect. When Abraham reprised, once again in print, his criticism of Einstein's theory, and concluded with the biting remark, "Einstein begs credit for the theory of relativity of tomorrow and appeals to his colleagues so that they may guarantee it,"[24] Einstein called it quits. Shortly after leaving Prague for Zurich in August 1912, he informed *Annalen* readers that he did not plan to respond to Abraham's latest arguments. "For the present, I would only like to ask the reader not to interpret my silence as agreement."[25] (Abraham's contention notwithstanding, Einstein never actually rejected the notion that the speed of light is the same in all gravitation-free frames of reference.[26]) Abraham remained unbowed. Throughout 1912, at various scientific meetings, he touted his own theory of gravitation and predicted the demise of Einstein's, whose original theory of relativity, he confidently assured an audience in Genoa, was fading away. "Will a new, more general principle of relativity arise like a phoenix from the ashes," he asked rhetorically "or will we return to absolute space?"[27] If his listeners had not yet caught on to the source of his hostility to Einstein's ideas, clarity came with his final question: "And will we bring back the much-disdained ether, so that it can take on not only the electromagnetic field, but also the gravitational field?"

In his public writings, Einstein did not hesitate to criticize Abraham's rival theory. In his private correspondence, he spoke his mind even more forcefully. "Abraham's theory has been created out of thin air, i.e., out of nothing but considerations of mathematical beauty,"[28] he wrote to one friend. "How this intelligent man could let himself be carried away with such superficiality is beyond me." In the same vein, he wrote to another colleague, "Though Abraham's new theory [a slight variation on an earlier version] is, as far as I can see, logically correct, it is nevertheless a monster spawned by embarrassment. The present relativity theory is surely not as wrong as Abraham says it is."[29]

Nevertheless, Einstein considered Abraham a first-rate physicist and recommended him for academic positions, including his own at ETH when he later moved to Berlin. A physicist with a sharp tongue himself, he seemed sympathetic to Abraham's barbed personality. "It is incomprehensible that this really important man is being avoided like the plague because of a few cocky sarcasms he indulged in a few years ago,"[30] he wrote in 1912 to the Swiss physicist and his own thesis adviser, Alfred Kleiner, who was looking to fill a vacant chair at the University of Zurich. Kleiner was unmoved. He couldn't abide Abraham nor could the search committee, which pointedly left his name off the list of candidates under consideration, without offering any reasons for the omission.

In 1913, Abraham became a full professor at the Milan Polytechnic, despite the efforts of some of his colleagues there to block his promotion. When rumors reached the ministry of instruction in Rome that Abraham's poor command of Italian prevented him from being an effective classroom teacher, the ministry held up the appointment pending a statement from the school's rector and a senator of the realm, Giuseppe Colombo. Coincidentally or not, Levi-Civita was the chair of Abraham's promotion commission, and he objected strenuously to the delay, pointing out that the Polytechnic's council of professors had approved the promotion without any reservations. In his letter to Colombo, Levi-Civita noted that Abraham's lithographed lecture notes on the theory of elasticity, an elective course he had offered for several years, had garnered high praise from his own colleagues. He added that anything short of immediate action to promote Abraham would, in his view, reflect poorly on the Polytechnic, inexcusably tarnish Abraham's reputation, and undermine the hard work of the promotion commission. As he informed Senator Colombo, "I have therefore declared that at the cost of remaining isolated, my vote would be resolutely contrary to any form of suspension [i.e., delay]."[31] Viewing the *sub rosa* objections to his friend's promotion through the prism of politics, Levi-Civita thought he discerned a more sinister motive. Amid rising tensions in Europe, which would ultimately see Italy take up arms alongside Britain and France against Germany in World War I, the first signs of the nationalism that would swallow Abraham up in 1915 had begun to appear on the campus. According to Levi-Civita, Milan's faculty saw in Abraham only a German citizen and was consumed by foolish talk, "perhaps disseminated and exaggerated by mistaken nationalism."[32] It is reported that when someone once asked him, "How are you getting along with your colleagues in Milan?" Abraham replied drily, "Excellent. I still haven't mastered the language entirely."[33]

Abraham was not the only scientist caught between the canon of classical mechanics and the changing face of physics in the early years of the twentieth century. Not long after Grossmann had initiated Einstein into the methods of the absolute differential calculus and they had produced their 1913 *Entwurf* paper outlining a generalized theory of relativity and a theory of gravitation, Einstein encountered varying degrees of resistance among his colleagues. Even Planck, who had been among the first wave of German physicists to embrace the special theory, had let slip to Wilhelm Wien, his coeditor at the *Annalen der Physik*, that Einstein's new theory of gravitation did not appeal to him. "The fraternity of physicists,"[34] Einstein wrote to his friend, the Swiss engineer Michele Besso,

> behaves rather passively with respect to my gravitation paper. Abraham seems to have the greatest understanding

> for it. To be sure, he fulminates against all relativity in *Scientia*... but he does it with understanding. I will visit Lorentz this spring in order to discuss the matter with him. He is greatly interested in it, like Langevin. Laue is not open to the fundamental considerations, neither is Planck, while Sommerfeld is more likely to be so. A free, unprejudiced look is not at all characteristic of the (adult) Germans (blinders!).

In describing the general attitude of the scientists he knew, the French mathematical physicist Marcel Brillouin emphasized that taste, temperament, and habit would accelerate or retard the acceptance of relativity theory. Along with Abraham he objected to the theory, in part because it ignored the hallowed idea of the omnipresent ether. Special relativity encountered resistance among Italian physicists for the same reason. Augusto Righi, the Bolognese pioneer of wireless telegraphy and the dean of Italian experimentalists, was not convinced that Michelson's ether-drift experiment, if done differently, would have yielded a negative result. Quirino Majorana, an accomplished experimental physicist, inherited Righi's chair at the University of Bologna, along with the same anti-relativity bias. Turin's mathematical physicist Carlo Somigliana, who enjoyed a long and deep friendship with Volterra and Levi-Civita, also never joined the relativity parade. In postcards acknowledging the reprints of relativity articles that Levi-Civita had sent him, Somigliana labeled himself an "unconvinced conservative."[35] A classically trained mathematical physicist, Somigliana was "relentlessly critical of relativity and the new quantum physics, which he neither understood nor tried to understand,"[36] a colleague at Turin remembers. In his role as critic, he harped on mistakes in logic or physical reasoning and remained puzzled by what he saw as misplaced faith on the part of his colleagues—scientists who, like himself, had been exposed to a rigorous education in the exact sciences. His former Scuola Normale classmate Vito Volterra met his exacting standards for doing science. "He is a pure classicist,"[37] Somigliana noted, "who always remained faithful to the school of Betti and of Beltrami, in which he grew up. His mathematical physics, however rich in originality, is similar to Helmholtz's, Lord Kelvin's and Kirchhoff's. In his output, there is no trace of vectorial theories, relativistic theories, the absolute differential calculus, no forays into the quantum field or wave mechanics."

Between early 1915, when Levi-Civita began a correspondence with Einstein, and the close of 1918, on the cusp of his departure for the University of Rome, he published fourteen papers on the theory of general relativity—the most important appearing as notes in the Lincei's *Proceedings* and other journals of that era. If Volterra by virtue of his ever-increasing prominence in mathematical, political, and international circles had become "Mr. Italian Science," as many called him, Levi-Civita most certainly was Mr. Italian Relativity. In November 1914, six months before Italy entered World War I,

Einstein sent the Royal Prussian Academy of Sciences a lengthy and complicated paper on the formal foundation of the general theory of relativity. Its main objective, he noted in the introduction, "is to close the gap"[38] between the mathematical methods used and the physical results obtained in the *Entwurf* paper that he and Grossmann had published half a year before.

In a second paper published in 1914, Einstein and Grossmann used a variational method to derive the *Entwurf* field equations, following a suggestion made by their Zurich colleague, the mathematician Paul Bernays. In recounting Einstein's epic intellectual odyssey from special to general relativity, science historians Hanoch Gutfreund and Jürgen Renn describe how the variational approach works and then how Einstein and Grossmann used it. Using such a method, they note, involves "tracing the evolution of a single function known as the Lagrangian."[39] As part of this process, they add, "energy-momentum conservation emerges as a natural by-product." When they came to do their calculations for this second and final joint paper, Einstein and Grossmann "succeeded in finding a Lagrangian function from which the *Entwurf* field equations could be derived, and they observed that this function is invariant under transformations between coordinate systems specified solely by the requirement of energy-momentum conservation."

In Einstein's November 1914 paper, written after his move to Berlin, he used a more refined and generalized variational method to derive his non-generally covariant field equations. Far from slighting the mathematics, he devoted almost half of the paper to deriving the key theorems of Ricci and Levi-Civita's absolute differential calculus— "simple derivations of the basic laws," as he expressed it—to spare his readers the necessity of reading more taxing (to his way of thinking) mathematical texts. Before deriving the gravitational field equations with the aid of these theorems, Einstein again reiterated the position he had staked out in the *Entwurf* articles that these equations "cannot possibly be covariant in all *generality*."[40] When offprints of the paper arrived, he posted a copy to his most vociferous—if competent—critic, Max Abraham. Einstein seems to have assumed that his intellectual sparring partner would promptly weigh in with an opinion. Abraham, however, had other ideas. "Carissimo amico," he wrote in late February to Levi-Civita, "I found yesterday a letter from Einstein, in which he points out that in his theory there is relativity with respect to *any* motion *of the origin* of coordinates. (??) [Abraham's emphasis and punctuation.] He also asks my opinion about his new paper."[41]

Having read the article in question himself, Levi-Civita was in a position to understand Abraham's frustration at not being able to follow all of Einstein's reasoning. "Truthfully, I didn't understand on which hypotheses [Einstein's] new proof is based," Abraham's letter continued, before coming to what he clearly considered the crux of the problem: "Among all the possible invariants that could be used to construct the $H$ function [the Hamiltonian] he chooses in a rather arbitrary way the one that leads to his field equations." To spare Levi-Civita the trouble of enlightening him in a letter, Abraham proposed that the two get together in the near future. Abraham and Levi-Civita may have shared a common interest in relativity theory, but Abraham was aware that he could not match Levi-Civita's grasp of the mathematical language used in formulating general relativity.

Impatient for their discussion to begin, Levi-Civita urged Abraham to come to Padua, and the two physicists met there a few days later. Their conversations about Einstein's November 1914 paper may well have inspired Levi-Civita, who had had no previous communication with Einstein, to subsequently get in touch with him abcut what he saw as flaws in the paper. Einstein replied quickly; during that spring, the letters fairly flew between Padua and Berlin. In March alone, Einstein wrote Levi-Civita four letters; in April, another six; in May, the number declined to one. Unlike Levi-Civita, who does not seem to have discarded much, if any, of his scientific correspondence, Einstein at this stage of his life was apparently more than a little more casual about saving letters from colleagues. Only one of Levi-Civita's 1915 letters to him survives, in contrast to the eleven letters Einstein wrote that year to Levi-Civita. Some years after Levi-Civita's death, his widow, Libera, presented those letters to the Einstein Archives, then based at the Institute for Advanced Study in Princeton.

Lacking Levi-Civita's letters, his share of the correspondence has to be reconstructed from Einstein's replies, and opinions as to its significance vary widely. Einstein biographer Abraham Pais writes that "the mathematical details of the [November] 1914 paper are of no interest for the understanding of the evolution of the general theory,"[42] and skips over its content without comment. "[Levi-Civita] pointed out some technical errors in Einstein's handling of the tensor calculus," says Pais, and "Einstein was grateful for having these brought to this attention." Somigliana, who knew Levi-Civita well enough to have probably heard firsthand about the correspondence, gives more credit to Levi-Civita, saying that he made "an essential contribution to the theory by correcting the form of Einstein's gravitational tensor so as to eliminate a physically inadmissible consequence."[43] Cattani and De Maria, after studying the letters in detail, conclude that "one of the starting points of [Einstein's] growing dissatisfaction with his *Entwurf* theory and of his new interest in general covariance can be traced back to his 1915 correspondence with Levi-Civita."[44] And while science historian John Norton notes that Einstein stoutly defended his theory from Levi-Civita's pointed challenge to his "derivation of the covariance properties of his gravitation tensor,"[45] he is careful not to suggest a causal connection between the final field equations and Levi-Civita's objections. The link—if any—between their correspondence and Einstein's "conviction that introducing adapted systems was on the wrong track and that a more broad-reaching covariance, preferably a *general* covariance, must be required"[46] remains an open question.

Einstein's first letter, sent to Levi-Civita in Padua, set the tone of the lively dialogue between the two scientists.

> Highly esteemed Colleague,[47]
> By examining my paper so carefully, you are doing me a great favor. You can imagine how rarely someone delves independently and critically into this subject. I also cannot help admiring the uncommon assurance with which you make use of a language that is foreign to you. When I saw that you are directing your attack against the theory's most important proof, which I had won by the sweat of my brow, I was not a little alarmed, especially since I know that you have a much better command of these mathematical matters than I. Nevertheless, upon thorough consideration I do believe I can uphold my proof.

The issues they were debating with such exuberant *politesse* were basically mathematical: Levi-Civita had apparently challenged Einstein's derivation of the field equations and, in particular, Sauer notes, "the tensorial character of the left-hand side of the field equations for admissible coordinate transformations."[48] Einstein's eagerness to communicate with Levi-Civita, which comes through quite explicitly in his letters, is understandable: From his standpoint, Levi-Civita was one of the very few "professional colleagues" who really understood relativity theory, and perhaps more importantly, cared enough to think through its implications. Concluding his first letter in early March 1915, he wrote, "I would appreciate it very much if you would let me know your opinion after having reconsidered the matter. With cordial greetings I am your devoted A. Einstein."

Levi-Civita appears to have replied with alacrity since less than two weeks later, Einstein wrote him once again, this time urging him to abandon German:

> I shall be delighted if next time you write me in Italian. I spent over half a year in Italy as a young man and at that time I had the pleasure of visiting the charming little town of Padova [Padua], and even now I still enjoy being able to apply my modest knowledge of the Italian language. On the other hand, I could not muster up the courage to write you in Italian, because the results would be far too clumsy and unclear.[49]

Then the letter settled down to business: "I shall limit myself to responding to your last letter," in which, evidently, Levi-Civita had mounted a point-by-point critique of Einstein's new proof. Their correspondence turned on a particular mathematical problem, involving an object that arose from the variation of the gravitational Hamiltonian and was to be interpreted as a gravitation tensor although its tensorial properties were under dispute. Einstein tried to answer these objections; and he voiced hope, at the close of the letter, that the mathematician would not find any errors in his proof. Levi-Civita's reply elicited yet a third letter from Einstein on March 20, which began, "Dear Colleague, I have received your letter with the counter-proof of the tensorial character [of the gravitation tensor] which is based on the case where among the adapted coordinate systems employed are such for which the [the components of the metric tensor] are constant." Einstein apparently had not yet persuaded Levi-Civita.

He tried again following the receipt of a yet another letter from Padua. Einstein knew how to begin a letter in a manner likely to disarm its recipient before moving in to decisively refute his correspondent's point of view. The opening paragraph of his fourth letter reveals his considerable talent along these lines:

> Dear Colleague,[50]
>
> I have just received your letter of March 23 written in the so-familiar Italian, of which I have been deprived for so long. You can hardly imagine what a pleasure it gives me to receive such a genuine Italian letter. While reading it, the finest memories of my youth come alive. You also make it very nice in your letters: First you flatter me nicely, to prevent me from making a dour face upon reading your new objections.

Evidently Levi-Civita possessed the same skill. The body of Einstein's letter, as before, is concerned with countering his objections.

From Milan, meanwhile, Max Abraham scribbled a hasty postcard to his friend, thanking him for the hospitality at Padua and inquiring, "How is the correspondence going with Einstein?"[51] Levi-Civita apparently had lost no time in letting Abraham know about his new correspondent—who, in fact, had shifted into high gear. On April 2, Einstein wrote again:

> Your letter of the 28th of March was extraordinarily interesting for me. I had to ponder for one and a half days without interruption before it became clear to me how to reconcile your example with my proof. I enclose your letter so that I can refer to it without any inconvenience to you. Your deduction is entirely correct. [The gravitation tensor] does not have a tensor character with the infinitesimal transformation envisaged by you, even though the transformation follows from a justified coordinate system. Oddly enough, my proof is *not* refuted by this for the following reason: My proof fails exactly in that special case you have addressed.[52]

Having initially conceded some ground to Levi-Civita, Einstein then spent the rest of the letter demonstrating how and why his proof was valid for all finite transformations. "This consideration," he added, "suggests a modification of my covariance proof. . . ." The letters is signed, as usual, "With cordial greetings, yours very truly," and then Einstein added, "I never had a correspondence as interesting as this before. You should see how I always look forward to your letters." In a letter that same week to his friend Heinrich Zangger, a professor of forensic medicine at the University of Zurich, he was even more expansive, writing. "The theory of gravitation will not find its way into my colleagues' heads for a long while yet, no doubt. Only *one,* Levi-Civita... has probably grasped the main point completely, because he is familiar with the mathematics used; but he is seeking to tamper with one of the most important proofs in an incessant exchange of correspondence. Corresponding with him is unusually interesting; it is currently my favorite pastime."[53]

Perhaps, as expressed in these few lines, the intellectual affinity that had blossomed in the space of two months between these two complex and driven scientists, both at the height of their powers, is the very quality that renders their 1915 correspondence historically significant.

Einstein and Levi-Civita exchanged several more postcards in April before Levi-Civita saw eye to eye with Einstein on the proof of the tensor character of [the gravitation tensor]. In his letter to the mathematician on April 14, Einstein acknowledged the pleasurable receipt of an "approving card," and added, "When the occasion arises to repeat the [gravitation tensor] proof, I will gladly include the corrections I have learned from our memorable correspondence." After bemoaning the fact that few in their scientific circle would be interested in seeing such a proof in print—"It is odd how few colleagues feel the intrinsic need for a *precise* theory of relativity"[54]—Einstein continued in a more personal vein, "Unfortunately, the attitude of our nonprofessional fellow mortals is incomparably more peculiar. So it is doubly gratifying to have come to know more intimately a man such as you. I will make great efforts to change our professional acquaintance into a personal one—one more reason at last to cross the Alps again one day. It is to be hoped that our fatherlands will not rebel against each other as well!"

Italy's decision only a few weeks later, to take up arms against its onetime treaty partner, Austria-Hungary, temporarily ended Einstein's hopes of visiting Levi-Civita in Padua (they would finally meet for the first time six years later in Bologna).

World War I disrupted Max Abraham's plans as well. Personally harassed by the anti-German militancy of the Milanese students, who shut down his class, and denounced as a German spy in Mussolini's jingoistic newspaper, *Il Popolo d'Italia,* Abraham packed his bags and fled across the border to Lugano, Switzerland.[55] In his last postcard to Levi-Civita from Milan, he lamented that "very few colleagues truly dared to stand up and censure the conduct of the students" and denounced one of his engineering colleagues for "covering the rct in the Polytechnic with the Italian flag," before inquiring, "Nothing new on the neo-relativistic chessboard?"[56] No reply has been found; it was Abraham's misfortune—one of many in his short, troubled life—to have served as a catalyst for events in which he would play no further part. His departure from Italy in 1915 marked the end of his involvement with Einstein's ideas. With no job to turn to in Italy or

Switzerland, he returned to Germany in 1916, fulfilled his required military service, and joined the German radio and telegraph company *Telefunken Gesellschaft,* where he worked on problems of radio transmission. Even after the war, a permanent chair in Germany eluded his grasp. He spent two semesters at Stuttgart's Institute of Technology, finding much to lament in the country's new Weimar Republic. "The spirit of the German university has been hardly changed by the revolution; on the contrary, it has grown worse,"[57] he later wrote to Levi-Civita. "I have always the hope to return to Italy." This hope, too, came to nothing.

In fall 1921, after losing an appeal to be reinstated at Milan, Abraham reluctantly agreed to be considered for a chair of mathematics at the Technical University of Aachen. If his scientific adversaries in Germany did not oppose the appointment, he told Theodore von Kármán, the Hungarian-born mathematician and physicist who had proposed Abraham's name for the Aachen job, there still remained the issue of anti-Semitism.[58] Abraham underestimated the head of Aachen's Aerodynamic Institute, who promptly obtained letters of recommendation from Göttingen's David Hilbert and two of Berlin's scientific luminaries, Planck and Einstein. In response to von Kármán's description of Aachen's ideal candidate as "someone who knows how to apply mathematical methods in a variety of creative ways," Einstein described Abraham's qualifications as follows:

> Abraham is, without doubt, one of the most gifted and deserving of living scientists in the world of mathematical physics. The strong point of his gift lies not in the area of physical idea formation, but in mathematical construction. I consider the thought of making him a teacher of mathematics at an Institute of Technology a very happy one; for he unites the fineness of scientific abstraction, in a purely mathematical area, with a rich knowledge of applied mathematics.[59]

Abraham's appointment followed swiftly, but the respite was brief. Barely six months later, while traveling by train to Aachen, he collapsed and was carried off on a stretcher. The cause was a brain tumor, and after a half-year of agony he died that November in Munich.

* * *

In the fall of 1915, six months after his flurry of correspondence with Levi-Civita ended, Einstein took the final steps in formulating general relativity. In the space of four weeks, over the course of four lectures to the members of the Prussian Academy, Einstein "wrestled with a succession of tensors, equations, corrections, and updates,"[60] before presenting his final version of the gravitational equations on November 25. "You must not be cross with me,"[61] he began one letter to Arnold Sommerfeld from Berlin, three days later, "that I am answering your kind and interesting letter only today. But in the last month I had one of the most stimulating, exhausting times of my life, indeed also one of the most successful. I could not think of writing.... I realized that my existing gravitational field equations were entirely untenable."

He then described several discoveries that finally led him to discard the *Entwurf* equations. One involved the anomalous motion of Mercury's perihelion, which had puzzled astronomers for more than half a century—namely the small discrepancy in the perihelion precession of Mercury's orbit from the value predicted by Newton's equations. Based on his earlier gravitational equations, Einstein had gotten a value of eighteen seconds per century, rather than the universally accepted value of forty-three seconds. When that November he returned to the requirement of generally covariant field equations ("a demand from which I parted, though with a heavy heart, three years ago when I worked together with my friend Grossmann"[62]) and discovered that his equations now explained the anomaly perfectly "without any special hypothesis," his heart skipped a beat. "I was beside myself with joy and excitement for days,"[63] he later told another friend.

The Dutch physicist Wander Johannes de Haas, who had recently collaborated with Einstein on a gyromagnetic experiment, remembers how he recounted the experience of "feeling that something actually snapped in him"[64] when he saw that his calculations agreed with Mercury's unexplained data. Pais maintains this discovery "was...by far the strongest emotional experience in Einstein's scientific life, perhaps in all his life."[65] His new theory also led him to predict that the bending of light rays in the vicinity of the sun's gravitational field would be twice as strong as he had predicted eight years earlier at the start of his quest to formulate general relativity. His dawning realization that Newton's seventeenth-century law of universal gravitation was a first approximation of his twentieth-century formulation of general relativity only deepened Einstein's conviction that Nature had spoken to him. His calculation of Mercury's orbit, in Norton's words, "gave the theory its first convincing empirical success."[66]

"I lost trust in the field equations I had derived [in 1914]," Einstein told his audience of Prussian scientists in early November, now ready to admit "that all his elegant manipulation of covariance requirements and adapted coordinates did not lead him to a definite set of field equations."[67] As he explained in his letter to Sommerfeld, "The covariance considerations in my paper last year do not yield the Hamiltonian function $H$. When it is properly generalized, it permits an arbitrary $H$. From this it was demonstrated that covariance with respect to "adapted" coordinate systems was a flop...[and] I saw clearly that it was only through a link with general covariance theory, i.e., with Riemann's covariant, that a satisfactory solution could be found."[68] In their correspondence earlier that year, Levi-Civita had drilled down on the tensor character of Einstein's gravitational tensor in his "adapted" system. In the intervening months, Einstein had apparently improved his understanding of the tensor calculus. "Nobody who really grasped it [general relativity] can escape its charm," he told his listeners, "because it signifies a real triumph of the general differential calculus as founded by Gauss, Riemann, Christoffel, Ricci, and Levi-Civita."[69]

In his fourth and final Academy lecture that November, Einstein presented his final mathematical equations for general relativity. "Every physical theory that complies with the special theory of relativity," he told his audience, "*can by means of the absolute differential calculus be integrated into the system of general relativity theory* [emphasis added]—without the latter providing any criteria about the admissibility of such physical theory."[70] In his formal presentation of the theory in the *Annalen* in March 1916, Einstein generously acknowledged the mathematicians who had helped him in various ways to reach his goal of incorporating all forms of motion into his theory. He credited Minkowski for his role in recognizing "the formal equivalence of space coordinates and the time coordinate"[71] and Grossmann for his indispensable help in combing through the mathematical literature and in the actual search for the field equations. Of Ricci and Levi-Civita, he had this to say: "The mathematical tools that are necessary for general relativity were readily available in the 'absolute differential calculus,' which is based upon the research on non-Euclidean manifolds by Gauss, Riemann, and Christoffel, and which has been systematized by Ricci and Levi-Civita and has already been applied to problems of theoretical physics.

"The general laws of nature," he continued, "are to be expressed by equations that hold good for all systems of coordinates, that is, are covariant with respect to any substitutions whatever."[72] Ricci and Levi-Civita could not have said it better.

CHAPTER 12

# Parallel Displacements

On June 25, 1914, not long after his forty-first birthday, Tullio Levi-Civita married Libera Trevisani, *dottoressa in matematiche* and his former student, in the city of Ferrara, north of Bologna. Three days later, the Serbian nationalist Gavrilo Princip assassinated Archduke Franz Ferdinand of Austria and his wife, Duchess Sophie, as they rode in their carriage through the streets of Sarajevo, in Bosnia. One month later, Austria-Hungary declared war on Serbia, and within weeks one European country after another, entangled in a web of interlocking alliances, rushed to commit mayhem on the battlefield. In May 1915, Italy's leaders took their nation into "the war to end wars" on the side of the Allies, as the coalition headed by Britain, France, and Russia came to be called.

For Levi-Civita and his new bride, the assassination of the heir apparent to the Austro-Hungarian Empire, spelled the end of their Alpine honeymoon in Cortina d'Ampezzo, an Austrian mountain hamlet just across the Italian border. In a postcard they jointly sent to Libera's sisters five days after Austria's official declaration of war, Levi-Civita wrote, "The international situation, and more precisely the risk of remaining with a mail and telegraph blackout, pushed us to return to Italy. Nina [Libera's nickname] withstood the journey (first in a wretched car service and then by train) with no problems. Let's hope that the warmer climate will accelerate her recovery."[1] (Libera was still recovering from an earlier surgery that had left her unable to have children.) The couple planned to stay a while in Belluno, a small town north of Venice, before making any plans, because, as Tullio explained rather sheepishly, "up to now, we have never managed to do anything right."

Italians remained deeply divided over entering the First World War. Those who called themselves interventionists included young students, galvanized by varying degrees of idealism and nationalism, the king, who had been brought up as a soldier, and industrialists who saw the war as a "source of new markets and new profits and as a way to weaken working-class organizations."[2] Others in the financial and manufacturing sectors refrained from taking a pro-interventionist stand, banking on the profits to be made from selling to both sides of the conflict. Political progressives and idealists saw the war as a battle between Prussian militarism and Italy's commitment to liberty and democracy, while Italian nationalists clamored for the return of the Trentino and Trieste (the "unredeemed lands" then held by Austria), which would give Italy a more natural Alpine frontier and control of the

Adriatic. Those who favored joining the Allies were in the minority—most working-class Italians and the peasants, who would make up half the army, rejected their patriotic exhortations—but "in the minds of [prime minister] Salandra and his foreign minister... none of these diverse currents of public opinion mattered much," notes a leading historian of the period.[3]

The charged atmosphere over Italy's role in the war permeated Levi-Civita's life in Padua, where he and his new bride had domiciled under the same roof as his parents. On one occasion, so the story goes, father and son were sitting at the dining-room table heatedly talking politics. Suddenly, Tullio, a socialist and an ardent pacifist, exclaimed loudly, "No, we must not go to war!"[4] His father, a veteran of the nineteenth-century struggle for Italian independence, who viewed the current conflict as its sequel with the laudable aim of recovering Austrian territories he considered Italian, angrily objected. "Be silent, Tullio!" At that point Libera interjected rather forcefully, "Let him speak, give him the floor!" Turning to his new daughter-in-law, Giacomo said, "Libera, either you apologize or leave this room." The strong-minded young woman promptly got up from the table and left the room.

"I remember a cultivated woman, beautiful, elegant, reserved,"[5] Susanna Silberstein Ceccherini, Libera's adopted daughter, would later recall of her mother. Libera was fiercely independent, an accomplished alpine hiker, and a good deal taller than Tullio. Her father, Luigi Trevisani, was born in Verona in 1849 and studied philosophy in Vienna, before returning to his hometown to teach moral philosophy and pedagogy at the local seminary, where he soon found himself at odds with the bishop over his refusal to take daily communion with his seminary students. He married Maria Speranza Scolari, a French teacher, in 1887, and went on to a successful teaching career in Italy's public high schools. Libera, the first of five sisters, was born three years later. A gifted student, she enrolled as an undergraduate in mathematics at the University of Padua in 1908 and received her degree *summa cum laude* four years later. One of only four women pursuing a degree in mathematics at Padua at the time, Libera filled her weekly schedule of classes with almost as many math electives as required courses, in fields ranging from the calculus of probability to geometric applications of the calculus. The history of art is also listed on her academic transcript—apparently the only humanities course she permitted herself in four years of study. In November 1912, she obtained a perfect score (30/30) on the *magistero in matematica,* the qualifying exam for teaching in the schools.

Her father, who embraced the then-radical idea that women deserved the same educational opportunities as men, was also not above exercising a certain patriarchal authority. Family lore has it that he gathered his daughters together before sending them off to university and decreed, "One of you cannot study mathematics." Pointing to Cornelia, the second youngest, he commanded, "you will study philosophy." Poor Cornelia went through life lamenting, "I wanted to study mathematics."[6] (Levi-Civita's sister, Ida, by way of contrast, was educated entirely at home, rarely rose before mid-morning, and seems to have spent her time embroidering tablecloths, tea towels, and napkins until the family arranged a suitable match for her.[7])

When Tullio visited Libera's home for the first time, most likely in 1913, to ask Luigi for his daughter's hand in marriage, Libera seemed taken aback. Her suitor was forty-one; she was twenty-four. Although she had written and defended her thesis for her mathematics degree on a problem in celestial mechanics under Levi-Civita's supervision, and published a paper on her subject as well, her contact with her academic mentor had otherwise been fairly limited. Seeking to reassure her, Libera's father stressed the benefits of marrying into a distinguished and wealthy Paduan family; and given Luigi's unorthodox ideas about organized religion, it is not likely that he saw any problem with his nominally Catholic daughter marrying a mathematician with a Jewish surname. He died in 1914, five months before the couple's civil marriage took place.

"You will be my companion for life,"[8] Tullio assured Libera after the marriage, and she lived up to her part of the bargain. Tullio's needs were few: he asked Libera to learn English and to master driving a car (she did both). And although Libera had once dreamed of teaching mathematics, she settled for traveling the world with Tullio, whose lecture schedule around the globe kept her busy. While he went to meetings and lectured, Libera attended plays and concerts and visited museums with her devoted friend Marcella Treves, who accompanied the couple on these trips. Cornelia, the sister resigned to studying philosophy, never married, taught literature in a junior high school in Rome, and lived with Libera and Tullio at their home in the capital. "They always lived together, because Tullio Levi-Civita worked constantly, so he was happy that my mother had company,"[9] according to family lore. Unable to have children herself, Libera suggested adopting a child, an idea that Tullio rebuffed. He seems to have had no interest in joining Libera on any of her cultural excursions, nor did he share her enjoyment of music, which he considered an obstacle to formulating ideas.

In autumn 1916, about a year after finishing his correspondence with Einstein, Levi-Civita submitted a paper entitled "Parallelism in Generic Manifolds and the Geometric Specification of Riemannian Curvature" to the *Proceedings of the Mathematics Society of Palermo* (*Rendiconti del Circolo Matematico di Palermo*), the internationally circulated journal of Italy's oldest mathematical society. His treatment in that paper of parallel displacement is today widely recognized by historians of science as "a landmark in the history of both general relativity and differential geometry."[10] In a letter about it to the Harvard mathematician George Birkhoff, he wrote, "I have continued to work on Einstein's theory," adding, "For the time being, all that I managed to do was to prepare a preliminary differential geometry memoir on the Riemannian curvature of manifolds: you know that it is precisely this curvature that plays a major role in the laws of reciprocal actions between physical phenomena and the surrounding space."[11] Between submarine warfare in the North Atlantic and America's decision to enter the war on the side of the Allies in the spring of 1917, Levi-Civita's letter announcing his discovery[12] took three months to reach Birkhoff in Cambridge, Massachusetts.

Levi-Civita had a modest goal in mind when he began studying this topic. As he reminded readers in the memoir's introduction, "the mathematical unfolding of Einstein's great idea (that finds in Ricci's absolute differential calculus its natural algorithmic instrument) intervenes as the curvature of a certain manifold in four dimensions and Riemann's related symbols." The driving force for him, initially, was "to reduce somewhat the formal apparatus that normally serves to introduce [Riemann's symbols] and to establish their covariant behavior."[13] The scope of Levi-Civita's investigation gradually enlarged, as he went on to explain: "Although this work was originally done with this sole motivation [the simplification of the Riemann curvature tensor], in its subsequent expansion it ended up giving a proper place to the geometrical interpretation."[14] Guido Castelnuovo, who would eventually stop working in algebraic geometry and explore other fields, including general relativity, was one of many on his colleague's mailing list. "Thank you for your papers," he wrote Levi-Civita. "I had seen [your articles] dedicated to relativity [and] have now read with much pleasure the Note on parallelism that you spoke to me about.... The simplicity of the results ensures that you are dealing with a *natural* concept in differential geometry, and it is strange that experts in this branch of mathematics had missed it until now."[15]

At the outset, Levi-Civita had turned to Riemann's writings for inspiration. "The main content of the paper, however, was devoted to the geometrical interpretation that he had looked for in vain in Riemann's papers,"[16] Bottazzini has pointed out. Following that unsuccessful exercise, the undaunted mathematician pressed on, first obtaining "the new derivation of the Riemann tensor, and then [seeking] a geometric interpretation of this derivation."[17] Indeed, a large portion of Levi-Civita's paper, which appeared in 1917, is given over to his search for a simple geometric interpretation of the Riemann curvature tensor. "Everybody knows how to move vectors on a plane,"[18] the mathematical physicist George Rainich wrote in his review of *The Absolute Differential Calculus,* Levi-Civita's classic work on the tensor calculus, referring to the Paduan mathematician's systematic use of parallel displacement throughout the text. Expanding on his explanation of how it works, Rainich continued,

> It is easy to extend the notion to surfaces which are applicable on a plane, that is, developable surfaces. In the case of a general surface, when he wants to displace a sector along a curve, Levi-Civita displaces it in the developable surface which is tangent to the given surface along

> the curve; this is his *parallel displacement.* He finds an analytic expression for this displacement and thus arrives at the required expressions with a simple law of transformation; these expressions are called the covariant derivatives.[19]

As one of Levi-Civita's students later observed about the discovery of the notion of parallelism: "The basic idea was simple, but its influence on the development of differential geometry was profound."[20]

In 1918, the Dutch mathematician Jan Arnoldus Schouten, a contemporary of Levi-Civita and, like Ricci, an early differential geometer, published his own formulation of a geometrical interpretation of the covariant derivative, using geodesically moving systems. Dirk Struik, then an assistant to Schouten at the Technical University in Delft, recalls how "Schouten came running into my office one day waving a reprint he had just received of Levi-Civita's paper. 'He has it too!' he cried out."[21] There were, to be sure, differences in their approaches, adds Struik: "Schouten's method was entirely intrinsic, whereas that of Levi-Civita utilized a surface imbedded in space. He also had derived his result only for case $n = 2$, although it was clear that it was intrinsic and valid for all values of $n$. The main difference... was that Levi-Civita's text was elegant and employed his absolute differential calculus ... whereas Schouten's work was difficult to read due to its unfamiliar notation."[22] For all these reasons, as well as the fact that Levi-Civita published his results a year before his Dutch colleague, his name is generally associated with the introduction of the theory of parallelism of vectors in a general Riemannian manifold.

Schouten might have been miffed at missing out on the recognition accorded Levi-Civita, but his admiration for Ricci, by now widely recognized for his contributions to general relativity theory, was boundless. He was to dedicate his first book, on the methods of the tensor calculus, written in German and published in 1924, to Ricci on the occasion of his seventieth birthday, hailing him as the founder of the absolute differential calculus. Thirty years later, when the name "Ricci calculus" had all but disappeared from the mathematics lexicon, Schouten persisted in naming his newest book *Ricci-Calculus* to commemorate "the memory of Dr. Gregorio Ricci Curbastro in life Professor of Mathematics in the University of Padua, who laid the foundations of tensor calculus."[23]

The years leading up to World War I had not been easy ones for Ricci. His beloved wife, Bianca, had died early in 1914, a loss that may have been partly ameliorated by the fact that her passing brought his daughter and her family back into his life. The Ricci matriarch, who ruled the household with a firm hand, had objected violently to the idea of her only daughter, Livia, marrying Remigio Lanzoni, an accomplished musician from Sant'Agata. The son of a local baker with anti-monarchist views, Remigio had been young Livia's piano teacher, and Bianca, who seems to have to put great stock in social standing, especially her own, would not countenance her daughter's union with someone from such a humble—not to mention, vaguely subversive—background. Despite her remonstrations, the couple did marry and afterward hastily departed for Egypt, where Lanzoni, a graduate of the Music Conservatory of Bologna, had established a flourishing career teaching, performing, and offering private lessons. According to family and local historians, mother and daughter never spoke again. The record is silent on Gregorio's role, but since Livia and her family only came back to Italy for visits after Bianca died, it seems likely that he resigned himself to the unhappy situation.[24] Following Bianca's death, he never remarried.

Italy's entry into World War saw Ricci's sons Giorgio and Cesare serving in separate regiments on the Italian front—four hundred miles of trenches dug in along the Alpine border with Austria, where the battles for the "unredeemed lands" of Trieste and Trentino took place. Ricci's public role during this period included serving as Padua's vice-mayor, a member of the city council, and alderman for education. To his sons, he was a devoted correspondent, although his letters, addressed to "the war zone" did not always reach their destination. When Giorgio, now a second lieutenant, intimated to his father that he wrote infrequently, Ricci, ever the mathematician, started numbering his postcards and letters sequentially to prove that the problem lay with Italy's postal system,[25] not his negligence. If he received an affectionate greeting, but no word about the sender's health, Ricci fretted for days. He seems to have spent a fair amount of time locating books, shoes, warm clothing, and food for his sons, packing them up securely and figuring out how to send them to their regiments, a task that became increasingly difficult as the war on the Austrian-Italian front dragged on. He invoked the blessings of Saint Anthony, Padua's patron saint, in his letters, imploring him to safeguard Cesare and Giorgio, and sometimes also enclosed prayer texts, which he urged his sons to use as well.

In winter 1916, Cesare was given a nine-day furlough. He visited his father in Padua and then continued on to Ravenna, to see his fiancée, who greeted him with a litany of complaints about what she evidently perceived as his negligent treatment. In response to the letter he sent his father afterwards, Cesare received a kind, but no-nonsense reply: "Shortly after your arrival here,"[26] Ricci began,

> I asked you if you were always happy with your fiancée and you told me you were; I have no reason to doubt the sincerity of your answer. On the other hand, the reproach addressed to you by Maria is not a sufficient reason for you to change your feelings towards her, since it is only proof of her legitimate desire to be sure of your affection, as I'm sure you want to be certain of hers. The misunderstanding might spring from the fact that you don't know the woman's heart well enough. A woman feels strongly

> the need to be able to rely on her partner's heart and support, while a man's love for a woman often takes a different form. In any case, the possibility for this cloud to lift depends on how you respond to your fiancée's complaints. If it's not too late, I would advise you to respond in the warmest way, which is what your heart will have advised you to do spontaneously, if you really love her.

Cesare's reaction to his father's advice is unrecorded, but in any case Maria soon vanished from his life.[27]

In June 1916, Cesare's regiment was involved in a fierce battle against an Austrian offensive at Monte Cengio, a critical Italian outpost in the Trentino. Taken prisoner, he was sent to a huge prisoner of war camp in Sigmundsherberg, Austria. He apparently spent the duration of the war there, a situation Ricci described in a subsequent letter to Giorgio as "certainly ill-equipped in some respects, but far from uncomfortable."[28] That fall, as fierce fighting north of Venice put Padua in the cross-hairs of the enemy's artillery, and fearing that the city might be occupied by Austrian troops, Ricci fled to Lugo, leaving his house in the care of his household staff, trusting them to find others to guard it if they left the city as well. By then, Padua had become the new command post of the Italian army. Temporary hospitals sprang up throughout the city, filling up rapidly with injured soldiers, while bombs began to rain down on the civilian population. In an effort to boost morale, King Victor Emmanuel III, in his role as head of state and nominal commander in chief, took up residence in a handsome villa on the periphery of the city.

In January 1918, Ricci, whose home in Padua had now become an officers' headquarters, went to Rome for a meeting at the Lincei. Later that year, Levi-Civita also headed back to the capital, intent on scouting out the possibility of transferring from Padua to the University of Rome. A decision in his favor, he knew, depended crucially on the support of Vito Volterra, head of the Roman school of mathematics. Would the fifty-five-year-old Volterra, an impassioned interventionist who had volunteered for military service in 1915, been commissioned as a captain in the Italian army and cited for meritorious service to the state in wartime, stand in his way? "I have a vague memory that my mother mentioned to me that Tullio faced a certain difficulty in obtaining the transfer to Rome because he was an avowed pacifist and against Italy's intervention,"[29] Libera's daughter, Susanna, later recalled.

That November, however Rome's tenured science faculty voted unanimously (13–0) to approve the transfer.[30] For the next twenty years, Levi-Civita remained at Rome, until the Fascist racial laws enacted in the summer of 1938 barred Jewish students from attending public schools and universities, Jewish authors from publishing works under their own names, and scores of Jewish academics, including some of Italy's best and brightest mathematicians, from teaching.

After taking up his new position at Rome, Levi-Civita joined forces with his fellow mathematicians to make Einstein's ideas more accessible, and in some cases more palatable, to the wider scientific community. Not only did he help to organize the first series of seminar talks on general relativity at the university, he also gave the first one, in March, 1918, with the provocative title "How can a properly conservative scientist bring himself to the threshold of the new mechanics?"[31] In his presentation, Levi-Civita built on variational methods in classical mechanics and generalizations from well-known formulas of Hamiltonian mechanics to reach Einstein's theory of gravitation, using relatively simple mathematics. His talk was really aimed at his longtime colleague Carlo Somigliana, who remained steadfastly opposed to the theory. Prefacing the mathematical part of his presentation, Levi-Civita remarked that no scientist need be fearful of the new, but that many were justified in taking a cautious approach to their research for their mission involved "zealously protecting" the established intellectual traditions and examining with "severe critical spirit" anything that modified or threatened that patrimony.

Three years after World War I ended, Einstein and Levi-Civita finally met face to face. The setting, fittingly enough, was Bologna, site of Europe's oldest university, where Einstein had accepted Federigo Enriques's invitation to lecture on relativity theory. During the third week of October 1921, he presented three public lectures to overflow crowds in the Archiginnasio, once the main building of the thousand-year-old university. The Italian newspapers covered his visit in exhaustive detail, publishing interviews with both Levi-Civita and Enriques, outlining the history of the theory, and noting with approval that Einstein gave his lectures in Italian (with a German translator nearby). In the pages of *Il Messaggero,* the public learned for the first time of the Einstein-Levi-Civita correspondence.[32]

Einstein then took the train to Padua, where Ricci, who had not been able to attend the talks in Bologna, had arranged for him to deliver a fourth lecture on October 27, 1921. There, in the university's storied Aula Magna, near where Galileo, three centuries earlier, had stood and lectured, the creator of general relativity and the mathematician who had invented its vital mathematical scaffolding met for the first and only time. Introduced by Ricci, Einstein, "spoke for an hour, slowly, accurately, in Italian, with a scientific precision," that, according to *Il Veneto,* the local newspaper, was amplified by "the linguistic accuracy of the speaker."

Although many people have searched for them, no photographs of Einstein with either Ricci or Levi-Civita have come to light. In a letter to Levi-Civita, written several days after his historic meeting with the physicist who had given the absolute differential calculus a new lease on life, Ricci marveled at "the opportunity to get close to him, more so, perhaps than I would have had in Bologna,"[33] and above all, "to admire the scientist and also the man, whose appearance reveals the height of his genius while emanating a personality unique to him." He must have derived as much if not more enjoyment from hearing Einstein say that he wished especially to express his great pleasure in presenting relativity theory in the very city where the architect of the absolute differential calculus taught.[34]

By this time, Italy's mathematical *doyens* had begun, however belatedly, to express new appreciation for the value of Ricci's work. Padua's Academy of Science, Letters and Arts led the way in overdue recognition, bestowing its highest accolade— "permanent" membership—on the sixty-three-year-old mathematician in 1915, after permitting him to languish for ten years as a corresponding academician. In Rome, the Lincei, which had elected Ricci a corresponding member in 1899 but had twice denied him the Royal Prize in Mathematics, suddenly promoted him to the rank of *Socio Nazionale* ("national member," its highest level of membership) in 1916, granting him the same status as Bianchi, Levi-Civita, and Volterra. The Venetian Institute of Sciences, Letters and the Arts, the first learned society Ricci was invited to join in 1892, and which had published many of his papers, elected him president in 1916. A number of similar organizations welcomed him into their ranks after World War I, including Turin's Royal Academy of Sciences in 1918, the Italian Academy of the Sciences—also known as the Academy of the XL for limiting its membership to a select forty souls—in 1921, and the Pontifical Academy of Sciences in 1925, shortly before Ricci's death. The city council of Padua, of which Ricci was a member, took the occasion of his sudden fame after Einstein's talk to acclaim him for his invaluable contributions[35] to a theory that in all likelihood few if any of them understood. Public praise from Einstein was more than enough to anoint him as a local hero.

These plaudits notwithstanding, "tired and melancholy,"[36] is how Ricci seemed to Francesco Tricomi, a member of Padua's mathematics department, in 1922. In his opinion, the department had seen better days. Many of its preeminent faculty, following in Levi-Civita's footsteps, were decamping for the University of Rome; many of those who remained behind were, like Ricci, nearing retirement age. That same year, Ricci's favorite grandson, Elio, who had been living with him while his parents were in Egypt, died at the age of nine, succumbing quite suddenly to some malady in those pre-antibiotic days. The child had "brought him comfort and consolation in somber years,"[37] Ricci later said.

Unlike Levi-Civita, whose mathematical oeuvre had by the end of World War I cemented his role as Italy's leading expert on general relativity, Ricci continued to focus his research on the intrinsic geometry of manifolds, starting with three-dimensional manifolds. "We speak of simply stated, well-defined problems for which a formulation by ordinary methods presents itself with dismaying complexity," Levi-Civita later remarked of this work.[38] Ricci, he added, "managed with subtle analysis to classify and resolve them in exemplary style."

Ricci had shifted his attention to applying his calculus to geometrical problems even before Einstein and Grossmann in their 1913 *Entwurf* paper had formulated the gravitational tensor (today called the Ricci curvature tensor) from arguments of covariance, publishing not only a book on this branch of mathematics, but also about two dozen articles between 1901 and 1925. Always interested in mathematical physics, he was long accustomed to using mathematics in the classroom to bring more rigor to physics, while simultaneously drawing attention to mathematics problems rooted in classical physics. Although he continued teaching Padua's mathematical physics course, his research now focused on using his calculus to extend Gauss's treatment of two-dimensional surfaces to arbitrary n-dimensional Riemannian manifolds.

As early as 1904, Ricci had published a short paper in which he used the absolute differential calculus to define his *Ricci curvature,* and then proceeded to use this tensor to define the principal directions and invariants of a manifold.[39] Roughly speaking, he hoped to use his tensor on arbitrary Riemannian manifolds in a manner similar to Gauss's application of his second fundamental form to arbitrary surfaces. He had limited success, according to the French differential geometer Marcel Berger, who in his definitive survey on differential geometry, wrote that "for Riemannian manifolds, Ricci was quite disappointed to find no geometric interpretation of the Ricci curvature lines."[40] Berger did not give a reference for this statement.[41]

More than a decade later, in 1917, Levi-Civita appears to have been the first to point out the close connection between Einstein's gravitational tensor and the Ricci curvature.[42] Some years later, in his classic text, *Lezioni di calcolo differenziale assoluto* ("Lessons on the Absolute Differential Calculus"), based on course lectures given at Rome in the early 1920s, Levi-Civita again called attention to the two tensors, this time linking them explicitly. Ricci, he said, had applied his tensor "to the study of the local distribution of curvatures in [a manifold of $n$-dimensions]," and "it was afterwards taken up by Einstein, who gave it a fundamental place in the theory of relativity (in which $n = 4$)." He added, perhaps ironically, "it is commonly known as the *Einstein tensor* [emphasis in original]."[43] It fell to one of Levi-Civita's younger colleagues at Rome, the differential geometer Enrico Bompiani, to formally name Ricci's tensor after the mathematician who originally introduced it, in 1922.[44]

Across the Atlantic, meanwhile, at Princeton University two of America's leading differential geometers, Oswald Veblen and Luther Eisenhart, had set their sights on transforming their university's differential geometry program into a world leader in the field. Eisenhart had become captivated by the subject as a Johns Hopkins University graduate student,[45] and never lost his deep affection for it. After writing his doctoral thesis on infinitesimal deformation of surfaces, he joined Princeton's mathematics faculty in 1900. Five years later, while traveling in Italy, he met Levi-Civita in Padua, who told him about Ricci's absolute calculus.[46]

Returning to Europe again in 1912, to attend the Fifth International Congress of Mathematicians in England, Eisenhart had a list of European mathematicians he and Veblen wanted to recruit for their faculty, including Castelnuovo and Levi-Civita, neither of whom, he reported ruefully, spoke English.[47] Castelnuovo never visited the United States; Levi-Civita had personal reasons for remaining in Padua at that time. When he finally visited Princeton in 1936 to give lectures and once again fraternize with Einstein (who had joined the Institute for Advanced Study there three years earlier) his command of English apparently still left a great deal to be desired. Leopold Infeld, then a visitor in Princeton's mathematics department, relates how he had been discussing an intractable problem in field theory with Einstein when Levi-Civita knocked on the physicist's office door. Einstein insisted he join the conversation, and Infeld listened silently while Einstein explained to Levi-Civita in English what he was working on. "I could not help laughing,"[48] Infeld later recounted in his autobiography, adding,

> Einstein's English was very simple, containing about three hundred words pronounced in a peculiar way. He had picked it up without having learned the language formally. But every word was understandable because of his quietness, slow tempo and the distinct, attractive sound of his voice. Levi-Civita's English was much worse, and the sense of his words melted in the Italian pronunciation and vivid gestures. Understanding was possible between us only because mathematicians hardly need words to understand each other. They have their symbols and a few technical terms which are recognizable even when deformed.

Presumably, when Eisenhart and Levi-Civita met in Padua in 1905 and again seven years later in Cambridge, they spoke to each other in French.

Eisenhart apparently introduced Ricci's methods in his graduate courses to solve problems in differential geometry as early as 1909. However, deformation of surfaces remained his chief research interest until general relativity burst upon the scene a few years later, bringing with it an explosion of interest and development in differential geometry. One of the first American mathematicians to delve deeply into Einstein's theory, Eisenhart also realized, perhaps to his surprise, that Ricci and Einstein's curvatures were the same, prompting Veblen to share his colleague's findings at a meeting of the National Academy of Sciences in Washington, D.C., in January 1922. Eisenhart's findings, quickly published as a brief note in the NAS proceedings, summed up in one succinct and understated paragraph his research findings, from start to finish. "In 1904," he began,

> Ricci developed the idea of principal directions in a Riemann space of $n$-dimensions, and in doing so introduced the contracted curvature tensor, which is fundamental in

> the Einstein theory, and gave a geometrical interpretation to it. A space in which these principal directions are completely indeterminate may be thought of as possessing a homogenous character. We derive Ricci's results by a slightly different method, and then show that the three types of space, chosen by Einstein in 1914, 1917, and 1919, as spaces free from matter are of this homogeneous character [i.e., the components of the Ricci tensor are all zero], and include all types of such spaces.[49]

Eisenhart introduced Ricci's life work to a much wider audience than the readers of Ricci's 1904 paper. A prolific book author who in 1926 published the first definitive volume in America on differential geometry, Eisenhart covered "tensor analysis in form and extent sufficient for the reader of the book who has not previously studied this subject,"[50] including an introduction to the geometrical interpretation of the Ricci tensor and parallel displacement. No longer regarded as an obscure branch of mathematics, the absolute differential calculus found its natural home in Eisenhart's book, which championed its methods from start to finish. In breathing new life into differential geometry, Eisenhart's classic work also put a premium on knowledge of Ricci's calculus. Hailed as "the first really readable Riemannian geometry,"[51] it remains in print to this day.

Ricci was familiar with Eisenhart's 1922 article. Two years after it appeared, he published a paper in the Lincei's *Proceedings* on manifolds with equal principal invariants, in which he pointed out in a footnote that the American mathematician "had highlighted the close relationship of this form [his tensor] with Einstein's gravitational tensor."[52] By then, he almost surely was aware of or had read for himself Bompiani's article on the geometry of curved spaces and Einstein's energy tensor in which he identified the Ricci curvature by the mathematician's name. Ricci's reticence on this occasion, if indeed he knew of Bompiani's gesture, is striking, but not inconsistent with his deeply ingrained modesty.

Despite the efforts of Eisenhart and Bompiani, the association of Ricci's name with his mathematical methods fell into disuse. When Levi-Civita published *Lezioni di calcolo differenziale assoluto* in 1925, its inventor's name figured prominently on the first page of the preface; but when the book's complete English translation was reissued nearly six decades later, the publisher, Dover Press, found it necessary to include the title *Calculus of Tensors* in parenthesis so as not to confuse its readers. Within two generations, any connection between Ricci's name and his absolute calculus had virtually vanished. On the topic of Einstein's gravitational tensor, Eisenstaedt had this to say: "The 'right' tensor, the one which is needed in the left member [of the field equations], is now called 'Einstein's tensor'—the least that one would expect. In fact, his tensor... is extremely similar to Ricci's... tensor... and coincides with it in the case of the vacuum. Not only did Einstein not obtain it from [David] Hilbert but, three years earlier, while he was working with Grossmann they had considered Ricci's tensor only to later abandon it, because they thought that with it they could neither recover Newton's equations nor satisfy any law of conservation of energy."[53] Both names are used in modern expositions of general relativity; the left-hand side of the field equations is identified as the Einstein tensor, which consists of two terms: the Ricci tensor proper, and its contraction (the Ricci scalar), multiplying the metric. As Eisenstaedt pointed out, the Einstein and Ricci tensor coincide in space-times with no matter, where all components of the two tensors are zero.

While Einstein and Grossmann did not, unlike Ricci, think of the Ricci tensor as geometrical, the geometrical interpretation later came to dominate the development of general relativity and differential geometry.[54] In mathematics, on the other hand, there do not appear to have been any significant applications of the Ricci curvature between 1904 and the early 1980s, when the mathematician Richard Hamilton, working toward a solution of Henri Poincaré's 1904 conjecture, "proposed a geometric evolution equation modeled after the heat equations of physics and named it 'Ricci flow' in honor of Gregorio Ricci-Curbastro, an early differential geometer."[55] Mathematicians who recognize Ricci's name today generally do so in connection with general relativity and the Poincaré conjecture. Nineteen twenty-four seems to have been the year that Ricci felt the need to correct the historical record. He may have been reluctant to put himself in the spotlight, but in another footnote in that same Lincei paper, he relates how, shortly after he published his first systematic account of covariant and contravariant differentiation in 1888, he told "the late Professor Ernesto Padova"[56] about his discovery of some differential identities satisfied by the Riemann tensor. Twelve years later, the differential geometer Luigi Bianchi rediscovered them independently, introduced them to his students at the University of Pisa, and published a proof using direct calculation (it would have been unthinkable for him to use the methods of Ricci's absolute calculus) the following year. He included his proof in the 1902 edition of his textbook on differential geometry and then turned his attention to other problems in the field.

The importance of the "Bianchi identities" and their relationship to Einstein's equations were recognized some years later, in the wake of general relativity. In particular, they mean that the field equations already imply the conservation of mass-energy-momentum.[57] Levi-Civita was among the tiny group of mathematicians, and certainly the first one in Italy, to grasp their significance. After years of neglect, the differential identities were "rediscovered" yet again by Levi-Civita, who, apparently unaware that Ricci had originated them thirty years earlier, introduced "several identities discovered by Bianchi" in the first of a steady stream of papers dedicated to general relativity, starting in 1917.

Toscano raises the question of why Ricci chose not to write out a proof and publish his discovery. Perhaps to a mathematician in 1888, he writes, these identities appeared "trivial, so trivial as to forget them."[58] It is also likely, adds Toscano, that Ricci did not find "anything significant in these relations, probably considering them a simple consequence of his formalism." Evidence that even Ricci had forgotten these identities can be found in an article he wrote on geodetic surfaces and published in 1903, one year after he heard Bianchi present his identifies at the Lincei. In preparation for using them in his own work, Ricci wrote, "Professor Bianchi... has called attention to several differential identities, which connect the symbols of Riemann that are encountered in the theory of the transformation of quadratic differential forms, with those of Christoffel." It is just possible that Ricci had not realized in 1903 that the differential relations established by Bianchi were identical to the differential relations he discovered thirteen years earlier using his absolute calculus. By 1924, however, he did, which may have been the motive for bringing up his past work. It is also likely that Levi-Civita, having learned for the first time the backstory from Ricci's 1924 paper, included it in his 1925 textbook and in its English translation[59] the following year.

Ricci died unexpectedly, following surgery, on August 26, 1925, at the age of 72. In his moving and emotional tribute, which he presented the following January in Rome at the Lincei Academy, Levi-Civita described how Ricci continued to teach, give exams, write papers, and organize his lectures for publication

> almost up to the day he entered a hospital in Bologna... having completed his teaching duties and enduring the prolonged stresses of the exam period, in which he participated against the explicit advice of his doctor... .The surgical procedure was seemingly successful: He was out of bed and looking forward to a recovery that would have allowed him to dedicate himself with renewed vigor to his research. But his heart wouldn't cooperate, and in a matter of hours a sudden attack of angina pectoris shattered his robust constitution.... He [had] requested a funeral without pomp, arranging that the family tomb in the cemetery of Lugo should display in his memory a simple tablet professing his profound devotion to his Catholic faith and summing up his entire life in this pronouncement: "He was for 45 years professor of mathematics at the University of Padua."[60]

In 1953, the hundredth anniversary of his birth year, the journal *Nature* published a commemorative essay that retraced the history of Ricci's calculus, recalling that

> it was so little thought of... that in 1901 Ricci was denied the Italian Royal Prize in mathematics on the ground that the calculus was 'useful but not essential for the treatment of some mathematical questions.' Nevertheless he himself

> retained a belief in its value, and in that had the sympathy and support of a young colleague at the University Padua, Tullio Levi-Civita.... Today... some knowledge of tensor analysis forms part of the ordinary intellectual equipment of most mathematicians and physicists... [and] the tensor calculus is fully established as one of the main instruments of modern mathematics and gives to its inventor a permanent place in the history of the subject. The verdict of 1901 no longer stands.[61]

Twenty years later, in December 1973, the Lincei celebrated the centennial of that "young colleague's" birth with three days of talks by leading mathematicians and scientists from the many fields in which Levi-Civita had worked. The Turin mathematician Tricomi, who knew him for many years ("I have said many times that, after my father and my closest relatives, Levi-Civita was one of the persons I loved most profoundly") took the floor before the program began "to insist on [his] exceptional human qualities"[62] and to emphasize that he deserved to be honored for his character and personal attributes, as well as his scientific achievements. In 1929 when Einstein proposed Levi-Civita for election as a corresponding member of the Prussian Academy of Sciences, he singled out for consideration two of Levi-Civita's accomplishments: his work with Ricci in developing the absolute differential calculus and his conception of parallel displacement.

Unlike Ricci, who died shortly after Benito Mussolini had transformed Italy into a Fascist dictatorship in 1925, Levi-Civita lived long enough to suffer its consequences. As *Il Duce* told the deputies in Parliament in spring 1925, "We are not a ministry; we are not even a government. We are a regime."[63] When the philosopher Benedetto Croce prepared an antifascist manifesto that spring, several hundred university professors, artists, and journalists signed it, including Levi-Civita. In all, more than one-fifth of the signatures Croce collected were those of Italian Jewish intellectuals.

In 1929, shortly after the regime's new state-sponsored cultural institution, the Reale Accademia d'Italia (Royal Academy of Italy), announced its first thirty members, the Vatican's Accademia Pontificia delle Scienze (Pontifical Academy of Sciences) elected Levi-Civita and Vito Volterra, also born into a Jewish family, to membership. Anti-Fascist organizations had been quick to note the absence of any Italian Jews among the first members of Mussolini's academy, and the Vatican was in this case equally quick to level the playing field. Disbanded in 1944, the Royal Academy during its brief existence was never to admit any Jews to its ranks.

In 1927, Mussolini's government announced that all professors within the national university system would be required to sign a loyalty oath professing allegiance to king and country, which Levi-Civita, along with nearly all his colleagues, duly signed. This seemingly innocuous pledge was followed in the fall of 1931 by a far more draconian order calling for allegiance to king, country, and the Fascist regime. Twelve of Italy's 1,250 university professors, including the elder statesman of Italy's mathematics community, Vito Volterra, refused to sign it and were consequently deprived of their teaching posts.[64] Others, like Levi-Civita and Giuseppe Levi, professor of anatomy at the University of Turin, delayed signing while they agonized over what they should do, short of resigning. In a letter to Levi-Civita that November, the Turin biologist declared that signing the new oath "would be an intolerable humiliation."[65] On November 19, Levi-Civita drafted (but did not send) a letter to the rector at Rome, Pietro de Francisci, inquiring whether he could assume that the new oath did not explicitly preclude him from dissenting spiritually from the political ideas of the regime. Nevertheless, both he and Levi signed the oath but not before Levi-Civita had sent a revised note to the university's rector stating that, as he saw it, the apolitical nature of mathematics made it unnecessary for him to declare his political ideals.[66]

Enacted in 1938, Italy's anti-Jewish legislation abruptly altered Levi-Civita's participation in the nation's intellectual life. Banished from the classroom and Rome's mathematics library, and expelled from numerous scientific academies and organizations, he "found himself cut off from all that made life interesting to him,"[67] the Scottish geometer Sir William V. D. Hodge wrote in his tribute to Levi-Civita in the obituary notices of London's Royal Society. According to Hodge, the mathematician learned of the decrees while staying with his sister... when someone happened to switch on the radio as they were being announced. His expression did not change, and he went out for his afternoon walk as usual." He suffered a further personal and professional blow when "the publisher of Springer-Verlag [Germany's major scientific publishing house] ordered its editor, Otto Neugebauer, to remove Levi-Civita's name from the masthead of the *Zentralblatt,* the renowned international review journal in mathematics, on the grounds that according to Italy's racial laws he was no longer a university professor."[68]

"The thing that left the strongest impression on my mother," Libera Levi-Civita's daughter said decades later, "is that many mathematicians, even those with whom there had been a close bond of friendship, had ceased all contact with Levi-Civita, as if they had never known him. This behavior was a hard blow for her and for Tullio, whose health began to totter just as this painful situation unfolded."[69] On December 29, 1941, three weeks after the attack on Pearl Harbor brought the United States into the war, Tullio Levi-Civita died in Rome at the age of 68. Although the Fascist regime prohibited Italian newspapers from announcing the death of Italian Jews, the Vatican newspaper *l'Osservatore Romano* published an obituary, hailing Levi-Civita's accomplishments as one of Italy's most distinguished mathematicians and pointedly noting his membership in the Pontifical Academy.

Speaking at a memorial service held in Milan after World War II ended, Carlo Somigliana, his colleague and friend for forty years, reflected on the pain and injustice inflicted on "a marvelous teacher and lecturer... whose clear and vibrant voice always delivered exquisitely precise thoughts."[70] His late friend, he added, "lived for science and the school [he created]—anything else he paid little attention to. When a stupid and inhumane measure deprived him of his chair, his life was cut short. Little by little, we saw him decline, until quite suddenly he faded away."

Following Levi-Civita's death, his widow, Libera, remained in Rome, where she died in 1973, at the age of 83, having spent much of her postwar life working to advance the cause of women's rights and education in Italy. In 1945 she adopted Susanna Silberstein, whose Jewish parents had placed her as an infant in a Florentine convent in 1943, shortly before they were rounded up by the Nazis and murdered in Auschwitz. Raised by Libera and Libera's extended family of sisters and friends, she came to know Levi-Civita and to honor his memory through the reminiscences and recollections of those who had known him well in his lifetime.

The story is told that Einstein, when asked what he best liked about Italy, replied, "Spaghetti and Levi-Civita."[71] It might just be true.

APPENDIX A

# From Ricci's absolute differential calculus to Einstein's theory of general relativity

MICHELE VALLISNERI

> The Absolute Differential Calculus is a conceptual, algorithmic theory, which allows the translation of the geometric and physical properties of space into an analytical form, independent of a choice of coordinates. It is, in its essence, the work of G. Ricci-Curbastro[...] It was applied to many problems of differential geometry and mathematical physics by Ricci and his disciples, including T. Levi-Civita; but it attracted the attention of all mathematicians when Einstein found in it an admirably complete, almost predestined tool for the mathematical development of his theory of general relativity. Since then the Absolute Differential Calculus has entered the common cultural estate of mathematics.
>
> Tullio Levi-Civita and Ugo Amaldi, "Absolute differential calculus," in *Enciclopedia Italiana* (Treccani, 1931)

Theoretical physics speaks the language of mathematics, yet the topics of the two disciplines could not be more different. Mathematics is fundamentally self-referential, concerned with the truth of its own constructs; physics is given the task of explaining and predicting the natural world. Thus, it is always a wonder[1] when a mathematical edifice born of logic and human ingenuity is revealed as the faithful mirror of a facet of reality. The achievement of this correspondence is seldom fortuitous; rather, it requires a courageous act of scientific imagination.

In 1913, Einstein and Grossmann performed such a feat by applying Ricci's absolute differential calculus (henceforth, ADC) to the formulation of a relativistic theory of gravitation.[2] Their *Entwurf* theory (as it came to be known) was flawed, for very interesting reasons that have been explored in great detail[3] and that I will briefly mention below, but its mathematical framework carried over essentially unchanged to the final theory of *general relativity* that Einstein completed in November 1915.[4] In fact, as a result of the *Entwurf* paper Einstein and Grossmann are credited with advancing mathematics as well as physics, by achieving a powerful synthesis of Ricci's calculus (itself based on foundational advances by Gauss, Riemann, Christoffel, Lie, and others) with the vector methods popularized in physical science by authors such as Hamilton, Grassmann, Gibbs, and Heaviside. Einstein and Grossmann named their formalism *tensor calculus*, extending the term *tensor*, insofar reserved for a few (second-order, symmetric) specific objects in mechanics and elasticity, to the wider and more abstract class of *systems of variables* explored by Ricci. Indeed, the second part of the 1913 *Entwurf* paper, which describes the tensor calculus, is regarded as the beginning of modern differential geometry.[5]

In this appendix, we shall briefly explore how Ricci's mathematical tools came to fit Einstein's vision, and how Einstein was able to imbue Ricci's formal mathematics with profound and innovative physical meaning. I will not be able to do justice to the breadth and depth of Einstein scholarship, which is unmatched across the history of physics. Rather, my outlook will be that of a relativist practicing in the 21st century, who contemplates the birth of his discipline with admiration, historical curiosity, and obviously with the benefit of hindsight. The mathematically minded reader interested in learning more details may begin from Zee[6] (for an intuitive path to the mathematics involved), Cooke[7] (for a mathematician's perspective), and the four-volume series on the inception of general relativity edited by Renn[8] (for a historical analysis).

Let me first attempt a concise statement, which I will then deconstruct by defining and clarifying mathematical terms and physical principles. In the years between 1905 and 1913, thanks to the work of Minkowski, Sommerfeld, and others, Einstein had learned to think of his theory of special relativity in geometrical terms:[9] he understood the geometrical *symmetries* and *invariants* of the *four-dimensional space-time* described by Minkowski, and he appreciated that the laws of mechanics and electromagnetism could be written as differential equations involving four-dimensional *vector* quantities. The special-relativistic equations of motion are then seen manifestly to maintain the same form when formulated from the point of view of any *inertial observer* (one of a privileged class selected by the symmetries of Minkowski space-time, as we will describe later). Thus, the equations of motion are said to be *invariant* with respect to transformations between the coordinates appropriate for different inertial observers, whereas the components of the vectors that appear in the equations change *covariantly*, in a well-ordered fashion ruled by the symmetries of Minkowski space-time.

Furthermore, Einstein had sought to generalize the laws of physics so that they would have the same form for *all observers in arbitrary states of motion*, including accelerated and rotating observers. Crucially, he had realized that any physical experiment witnessed by a uniformly accelerated observer would be indistinguishable from the same experiment as described by an inertial observer in the presence of a suitable homogeneous gravitational field, thus linking the extension of relativity to general observers with the task of formulating a theory of gravitation compatible with special relativity. This was Einstein's celebrated *principle of equivalence*, which explains Galileo's observation (that gravity, just like the fictitious *inertial* forces due to the acceleration of reference frames, acts equally on bodies of different masses and compositions) by *identifying* gravitational and inertial forces. Following this train of inquiry, Einstein had realized that acceleration (and therefore gravitation) implied that the speed of light would come to depend on location within a gravitational field, and had been inspired to describe this state of things by way of a *curved space-time* analogous to Gauss's curved surfaces.

General relativity, as well as the 1913 *Entwurf* theory, lies further along this direction. Space-time can be curved along all four dimensions; its geometry and curvature are described locally, *à la* Gauss, by a *line element* (also known as *metric tensor*) with ten independent components, rather than the variable speed of light alone. Moreover, space-time can be parametrized with any four smoothly varying coordinates; for a space-time with no symmetries, there will be no preferred coordinate system. As Grossmann showed Einstein once the two were reunited in Zurich, the problem of converting the special-relativistic equations of non-gravitational physics to a *generally covariant* form that is valid across curved space-time is solved by a straightforward but masterful application of the ADC. Following Ricci's method, one replaces ordinary derivatives with Ricci's *covariant derivative*, and postulates that all vectors (indeed, all *tensors*) transform as the ADC's covariant or contravariant systems of variables.

In the *Entwurf* paper, Grossmann emphasizes how the ADC can be applied without modification to describe flat manifolds using general coordinates (as in the case of the physical applications outlined in Ricci and Levi-Civita's 1900 review,[10]) or to treat *arbitrary curved manifolds*:

> The vector analysis of Euclidian space related to arbitrary curvilinear coordinates is formally identical to the vector analysis of an arbitrary manifold given through its line element. Therefore, there are no difficulties in extending the vector analytic conceptual system, as it had been developed in recent years by Minkowski, Sommerfeld, Laue, and others for relativity theory, to the general theory of Einstein given here.
>
> The *general vector analysis*, which one then recovers, proves with some practice to be just as simple to manage as the special vector analysis of three- or four-dimensional Euclidian spaces.

Thus, in 1913 Einstein and Grossmann had all the mathematics they needed. The remaining hurdle to a complete theory of gravitation was to formulate the field equations for the gravitational potentials, which Einstein had already identified correctly in the components of the metric tensor. However, the quest for the field equations would prove very troublesome. Subtle mistakes and misleading hints would continue to bedevil Einstein until November 1915,[11] when he finally converged onto his triumph: general relativity, a paragon of elegant mathematics applied to explain and predict Nature, and an unexpected vindication of Ricci's so far scarcely noticed masterpiece.

***

**Vectors, covariance, invariance.** Let us begin our exploration of terms and principles at a familiar place in elementary physics: the non-relativistic equations of motion for a massive particle attached to a spring, and confined to the plane:

$$ma^i = m\frac{d^2x^i}{dt^2} = F^i = -kx^i. \tag{1}$$

Here $x^i$ (with $i = 1, 2$) represents the particle's position, given as familiar Euclidian coordinates $x$ and $y$; $m$ is the mass of the particle and $k$ is the spring constant. In words, mass $\times$ acceleration $=$ restoring force, proportional to the extension of the spring.

Now, we know from analytical geometry that if we rotate the plane counterclockwise by the angle $\theta$, the coordinates $x$ and $y$ transform as

$$\begin{aligned}\tilde{x} &= \cos\theta\, x - \sin\theta\, y,\\ \tilde{y} &= \sin\theta\, x + \cos\theta\, y,\end{aligned} \tag{2}$$

where $\tilde{x}$ and $\tilde{y}$ are post-rotation coordinates. Any physical quantity consisting of two numbers that transform like the coordinates $x$ and $y$ is known as a *vector* (in this case, a two-dimensional Euclidian vector). That is manifestly the case for the acceleration and force in Eq. (1), since both are linear functions of the coordinate vector. Thus, under a rotation the equation as a whole transforms to

$$m\tilde{a}^i = m\frac{d^2\tilde{x}^i}{dt^2} = \tilde{F}^i = -k\tilde{x}^i. \tag{3}$$

In this case, we say that the equation is *covariant* with respect to rotations, since it involves mathematical objects that transform homogeneously according to the rotation law (2). In addition, the form of the equation, as well as the physics that it expresses, are said to be *invariant*: they are the same in any rotated frame.

Establishing (or postulating) the covariance of an equation poses a powerful constraint on the physics that it represents. For instance, non-vectors such as $(x^2y, xy^2)$ cannot appear in rotationally covariant equations. By contrast, these may feature the scalar product $\sum_i (x^i)^2 = x^2 + y^2$, which is itself *invariant* with respect to rotations.[12] Indeed, rotations (as well as reflections) may be defined as the set of transformations that leave scalar products invariant.

We can break the covariance of Eq. (1) by applying more general transformations, such as the transformation to a *rotating* coordinate frame:

$$\begin{aligned}\tilde{x} &= \cos\omega t\, x - \sin\omega t\, y,\\ \tilde{y} &= \sin\omega t\, x + \cos\omega t\, y,\end{aligned} \tag{4}$$

where $\omega$ is the angular rotation velocity. To recast (1) in terms of rotating-frame coordinates, we solve for $(x, y)$ in terms of $(\tilde{x}, \tilde{y})$, replace the resulting expressions in Eq. (1), take the time derivative, and then recombine the resulting equations as

$$m\frac{d^2\tilde{x}}{dt^2} = -k\tilde{x} + m\omega^2\tilde{x} - 2m\omega\frac{d\tilde{y}}{dt}, \tag{5}$$

$$m\frac{d^2\tilde{y}}{dt^2} = -k\tilde{y} + m\omega^2\tilde{y} + 2m\omega\frac{d\tilde{x}}{dt}, \tag{6}$$

or more compactly as

$$m\frac{d^2\tilde{x}^i}{dt^2} = -k\tilde{x}^i + m\omega^2\tilde{x}^i - 2m\omega\times\frac{d\tilde{x}^i}{dt}, \tag{7}$$

where the cross product "$\omega\times$" has the effect of multiplying a vector by $\omega$, and rotating it by 90°. This equation features two additional terms, described in elementary-physics textbooks as the centrifugal and Coriolis forces.[13] In other words, the physics described by Eq. (1) is not invariant with respect to transformations that take us to a rotating frame; new physical effects appear if we do that. Formally, we say that Eq. (1) is not covariant with respect to time-dependent rotations.

The new forces arise because the coordinate transformations (4) are time-dependent, so they interfere with the time derivative $d/dt$. It is possible to *expand the covariance* of Eq. (1) by replacing the time derivative $d/dt$ with the more general operation

$$\frac{D}{Dt} = \frac{d}{dt} + \omega\times, \tag{8}$$

which expresses the fact that in rotating frames all vectors are seen to rotate in addition to any other change due to physical interactions. Equation (1) is then written as

$$m\frac{D^2\tilde{x}^i}{Dt^2} = m\frac{D}{Dt}\frac{D}{Dt}\tilde{x}^i = -k\tilde{x}^i, \tag{9}$$

and in this form it is valid in any rotating frame, since the quantity $\frac{D^2\tilde{x}^i}{Dt^2}$ again transforms as a vector. The operation $D/Dt$ is a very simple example of *covariant derivative*, in a very limited context that greatly simplifies its mathematical implementation, and that yet exemplifies the theme of extending the covariance of a theory.

**Space-time in special relativity.** In pre-relativistic physics, three-dimensional Euclidian geometry provides a passive stage for all physical phenomena, which evolve according to the tics and tocs of a *universal time* shared by all systems and observers. One would then describe a solution for (say) an extended field as a mapping $\phi(\mathbf{x}, t)$ from space $\times$ time $(\mathbb{R}^3 \times \mathbb{R})$ to the space of field configurations.

Einstein founded his 1905 theory of special relativity[14] on a physically motivated critique of the notion of time, and specifically of *absolute simultaneity*, which becomes untenable when the propagation of physical effects is limited to the speed of light. Rather, *inertial observers* in relative motion with respect to each other come to have different definitions of simultaneity, corresponding to three-dimensional hypersurfaces that are *tilted in space-time*. These different definitions are not merely mathematical sophistications, but have very concrete and observable effects in the phenomena of time dilation and length contraction—startling predictions that would later be validated by countless experiments.

The preferred class of inertial observers includes all those that measure space and time as homogeneous, isotropic, and time-independent, and that witness no fictitious inertial forces due to acceleration. Any observer in constant linear motion with respect to an inertial observer will also be inertial. In special relativity, the notion of inertial observer is often identified with that of *inertial coordinate frame*, in that each inertial observer can be assigned coordinates that correspond naturally to his or her (notional) time and length measurements—not just locally, but across space and time. No such identification is possible, in general, for accelerated observers or in the presence of gravitational fields. Inertial coordinates are related by simple equations (the *Lorentz transformations*) that arise from symmetry arguments, together with the postulate that the speed of light is measured to be the same by all inertial observers.

Einstein's 1905 treatment is couched in these algebraic transformations, but the mixing of the space and time dimensions calls for a geometrical description in terms of a *four-dimensional space-time*, named after Hermann Minkowski.[15] Minkowski space-time can be identified with $\mathbb{R}^4$ endowed with additional structure: the *space-time interval* $-c^2\Delta t^2 + \Delta x^2 + \Delta y^2 + \Delta z^2$ (with $c$ the speed of light), which remains invariant in any inertial frame.[16] When negative, the space-time interval expresses the *proper time* measured by a clock moving across the interval; when positive, it expresses the *proper distance* measured by a ruler *at rest* in the inertial frame where the extrema of the interval are simultaneous.

Minkowski's justly famous quotation expresses the radical import of the notion of space-time, as evident in his reformulation of special relativity:

> Henceforth space by itself, and time by itself, are doomed to fade away into mere shadows, and only a kind of union of the two will preserve an independent reality.[17]

Just as important, in terms of the development of physics, was the new connection between the symmetries of space-time and the covariant formulation of physical laws. When expressed in terms of four-dimensional vectors, the equations of special relativity (which include the motion of point masses, but also the evolution of the electromagnetic field) are covariant with respect to Lorentz transformations. Conversely, the requirement posed by Einstein that the equations of physics have the same form according to all inertial observers (the *principle of relativity*) meant that only covariant equations would henceforth be acceptable in formulating a theory. This idea would come to play a major role in the genesis of Einstein's theory of gravitation, and more broadly for quantum field theory and many other disciplines in modern physics.

**Riemannian manifold, metric tensor, covariant derivative.** I will now introduce Ricci's covariant derivative more formally, in the specific context of Einstein's campaign to extend the theory of special relativity to include gravitation, which requires replacing Minkowski's space-time with a more general curved geometry. For this introduction, I will follow Ricci and Levi-Civita's (henceforth RLC) 1900 review[18] of the ADC. The review is surprisingly readable; many formulas will be instantly recognizable to modern readers familiar with differential geometry. Such readers may enjoy the review in Robert Hermann's 1975 translation,[19] which employs modernized notation and terminology, and offers extensive comments that link RLC's early synthesis to modern developments and applications.

RLC ground the ADC in Riemann's notion of *n-dimensional manifold* (in French, *varieté à n dimensions*). The geometry of Riemannian manifolds is defined *locally* and *intrinsically* by an equivalence class of quadratic differential forms, with two forms equivalent if they can be mapped into each other by a change of coordinates. It follows that manifolds are geometrically invariant with respect to general coordinate transformations.

The quadratic form that specifies local geometry is known as *first fundamental form* (or, now more commonly, as *line element* or *metric*). We write it as

$$\text{(10)} \qquad \mathrm{d}s^2 = \sum_{ij} g_{ij} \mathrm{d}x^i \mathrm{d}x^j, \quad \text{with } i,j = 1,\ldots,N;$$

it relates infinitesimal displacements in the coordinates $x^i$ to infinitesimal proper distances[20] $\mathrm{d}s$. A coordinate transformation[21] $x \to y(x)$ maps $g_{ij}$ into a numerically different $g'_{ij}$, but the line element remains invariant:[22]

$$\text{(11)} \qquad \mathrm{d}s^2 = \sum_{ij} g_{ij} \mathrm{d}x^i \mathrm{d}x^j = \sum_{ij} g'_{ij} \mathrm{d}y^i \mathrm{d}y^j,$$

where the transformed metric coefficients and coordinate displacements are given by

$$\text{(12)} \qquad g'_{ij} = \sum_{\ell m} g_{\ell m} \frac{\partial x^\ell}{\partial y^i} \frac{\partial x^m}{\partial y^j}, \quad \mathrm{d}y^i = \sum_{\ell} \mathrm{d}x^\ell \frac{\partial y^i}{\partial x^\ell}.$$

To visualize the difference between local/intrinsic and global/extrinsic geometrical properties, it is useful to think of a curved two-dimensional surface embedded in three-dimensional Euclidian space, as we may realize by deforming a sheet of rubber. The proverbial ant constrained to the surface may perform any geometric "experiment" having to do with measuring distances and angles, and she would find that its results are fully determined by the first fundamental form. However, were the ant to raise her eyes skyward, she may realize that the folding of the surface makes it possible to reach, with a single hop, a destination that she had laboriously measured to be many steps away. Such a global property, which relates to the embedding of the surface in higher-dimensional space, requires description by additional mathematical structure (such as the so-called *second* fundamental form); nevertheless, Riemannian manifolds can be defined purely in intrinsic terms, with no reference to or need for embedding. It turns out that Riemannian ants have no sky to look at, just as we have no outlet into a fourth spatial dimension.

The ADC, RLC continue, aims to extend the coordinate invariance of manifolds to laws and equations written in terms of *covariant* and *contravariant forms.* What are these? Let us again consider the coordinate transformation $x \to y(x)$. Under the transformation, the partial derivatives $f_i \equiv \frac{\partial f}{\partial x^i}$ of a function $f$ map to

$$f'_i \equiv \frac{\partial f}{\partial y^i} = \sum_\ell \frac{\partial f}{\partial x^\ell} \frac{\partial x^\ell}{\partial y^i} \quad \text{(covariant);} \tag{13}$$

any single-index object that transforms in the same fashion is called *covariant,* because it transforms in the "same direction" as the coordinate axes. Covariant objects with more than one index are possible, such as the coefficients $g_{ij}$ of the line element; to transform these, as we saw in Eq. (12), we multiply by a $\frac{\partial x}{\partial y}$ matrix for each index.

We have already seen that the differential forms $\mathrm{d}x^i$ transform to

$$\mathrm{d}y^i = \sum_\ell \mathrm{d}x^\ell \frac{\partial y^i}{\partial x^\ell} \quad \text{(contravariant),} \tag{14}$$

as do the coefficients $A^i$ in a function $g = \sum_i A^i \frac{\partial f}{\partial x^i}$; all these objects are called *contravariant.*

In our exposition we have used modern differential-geometry notation, writing lower indices to denote covariant quantities, and upper indices for coordinates and contravariant quantities. By contrast, RLC and Einstein–Grossmann adopt different inhomogeneous notations, informed by earlier contributions. In what follows, we will also omit summation signs whenever a sum occurs over paired lower and upper indices—a convention due to Einstein that carries his name.

The transformation laws of covariant and contravariant quantities help us reformulate mathematical and physical equations that involve simple partial derivatives of functions; however, objects formed with higher-order derivatives are neither covariant nor contravariant. What then? The difficulty is surmounted thanks to a "remark" (RLC's word) made by Christoffel in the context of the theory of differential invariants,[23] and elevated by Ricci to cornerstone of his calculus. Namely, if $X_i$ is covariant, then

$$X_{i;j} = \frac{\partial X_i}{\partial x^j} - \begin{Bmatrix} k \\ ij \end{Bmatrix} X_k \tag{15}$$

*is also covariant.* Here

$$\begin{Bmatrix} k \\ ij \end{Bmatrix} = g^{k\ell} \frac{1}{2} \left( \frac{\partial g_{i\ell}}{\partial x^j} + \frac{\partial g_{j\ell}}{\partial x^i} - \frac{\partial g_{ij}}{\partial x^\ell} \right) \tag{16}$$

is the *Christoffel symbol*,[24] with $g^{k\ell}$ the inverse metric, such that $g_{ik}g^{k\ell} = \delta_i^\ell$ (i.e., the diagonal identity matrix). Note our use of the Einstein summation convention.[25]

This new operation, denoted in modern notation by semicolon-plus-index, is the *covariant derivative* with respect to the metric, which (in RLC's words) "acts on covariant and contravariant forms to derive others of the same nature." It applies slightly differently to contravariant tensors ($X^i{}_{;j} = \frac{\partial X^i}{\partial x^j} + \left\{ {i \atop jk} \right\} X^k$), and it involves more terms when operating on objects with more indices, but it remains linear with respect to the object being differentiated.

Armed with the covariant derivative, RLC lay out the gist of their method:

> When one meets a new problem, to put the equations in *invariant form* [our emphasis] it suffices to express its main features in terms of general coordinates [using covariant and contravariant quantities], and then to substitute covariant differentiation for ordinary differentiation.[26]

The substitution is safe because the covariant derivative reduces to the ordinary kind in the Cartesian coordinate systems in which physical and geometrical problems are initially formulated; in these systems the Christoffel symbol is identically zero, or (for slightly more general systems) it yields simple intuitive expressions in the vein of Eq. (8).

Next, RLC define the Riemann curvature tensor,

$$R^\ell{}_{ijk} = \frac{\partial}{\partial x^j} \begin{Bmatrix} \ell \\ ik \end{Bmatrix} - \frac{\partial}{\partial x^k} \begin{Bmatrix} \ell \\ ij \end{Bmatrix} + \begin{Bmatrix} m \\ ik \end{Bmatrix} \begin{Bmatrix} \ell \\ mj \end{Bmatrix} - \begin{Bmatrix} m \\ ij \end{Bmatrix} \begin{Bmatrix} \ell \\ mk \end{Bmatrix}, \tag{17}$$

and prove the *Ricci identity* that quantifies the failure of covariant derivatives to commute in terms of the Riemann tensor:[27]

$$X_{k;ij} - X_{k;ji} = R^\ell{}_{kij} X_\ell. \tag{18}$$

For a *flat* metric (one that can be recast as $\delta_{ij} \mathrm{d}x^i \mathrm{d}x^j$ in suitable coordinates), the Riemann tensor is identically zero in all coordinate systems, so derivatives commute. The converse is also true.

**From special relativity to general-relativistic theories, by way of the ADC.** RLC have much more to discuss in their review, including several worked-out applications of the ADC. We, however, are now ready to make the jump to Einstein and Grossmann's (henceforth EG) *Entwurf.*

Where RLC can only claim that the ADC produces "elegance, agility, and insight," Grossmann sees opportunity for the new physics sought by Einstein. The laws of special relativity are covariant, but only with respect to the limited group of Lorentz transformations between inertial observers. By deploying the ADC, one may generalize these laws not just to arbitrary coordinates, but to *any Riemannian manifold.*[28] Doing so opens the space for the metric tensor $g_{ij}$ to play the dual role of line element for space-time and of *gravitational potential* sourced by mass/energy–momentum.

Let us exemplify the first, geometric role of the metric by considering the special-relativistic Maxwell equations,

$$\sum_j \frac{\partial F_{ij}}{\partial x^j} = J_i, \tag{19}$$

with $F_{ij}$ the electromagnetic field, and $J_i$ the electric charge–current four-vector. Replacing the ordinary derivative with the covariant derivative with respect to $g_{ij}$ elevates Eq. (19) to the generally covariant Einstein–Maxwell equations,

$$F_{ij}{}^{;j} = J_i, \tag{20}$$

which are valid in any coordinate system and in any manifold, whether the flat Minkowski space-time of special relativity, or the truly general-relativistic space-times curved by matter.

In curved space-times the Riemann tensor is not identically zero, so there is no coordinate transformation that brings the equations back to the special-relativistic form of Eq. (19). The best one can do is to find transformations that realize Eq. (19) in the neighborhood of a space-time point, materializing a freely-falling *local Lorentz frame* where the laws of special relativity apply unchanged.[29] Thus, the validity of special-relativistic equations elevated to curved space-time requires a greater leap of faith than RLC require of their readers, but it is a leap that carries a beautiful physical interpretation by way of Einstein's principle of equivalence, and that yields, in one elegant swoop, consistent laws for the interplay of gravitation and non-gravitational physics.

The second, dynamical role played by the metric is that of *gravitational potential.* By 1913, this was already clear to EG: in the *Entwurf*, the metric determines the inertial motion of *test* masses[30] along *geodesics* – the space-time trajectories that maximize proper time integrated along them – just as it does in the 1915 theory of general relativity. It also determines the evolution of continuous mass distributions, by way of a gloriously simple equation for the conservation of their energy–momentum:

$$T^{ij}{}_{;j} = 0. \tag{21}$$

Thanks to the ADC, these equations of inertial motion are covariant with respect to general coordinate transformations, a crucial goal of Einstein's program.

EG were instead uncertain about the form of the field equations that determine $g_{ij}$ from the evolving distribution of mass/energy–momentum. They were impressed by the fact that the Riemann tensor yields a "complete system of differential tensors" involving second derivatives of the metric (just what they expected to appear in the field equations); and they reckoned that

> the extraordinary importance of these [tensors] for the *differential geometry* of a manifold [...] makes it a priori probable that [they] may also be of importance for the problem of the differential equations of the gravitational field.[31]

This line of reasoning leads to generally covariant field equations written in terms of the *Ricci tensor* $R_{ij} = R^{\ell}{}_{i\ell j}$, as they would be in the final theory of general relativity.

However, the weight of empirical evidence dictated that EG's theory should also recover the Newtonian equations of gravitation in the limit of weak fields and slowly moving objects. Unfortunately, the specific form of the metric that EG assumed for that case, while phenomenologically adequate, turned out to be incompatible with the coordinate *restrictions*[32] under which they attempted to develop the field equations. EG's reasons for these choices have been parsed in great detail—a very technical topic, yet one with intriguing implications about EG's philosophical and methodological stance.[33]

As a result, EG had to resort to formulating field equations written in terms of non-fully covariant mathematical objects, and therefore characterized by limited covariance: to wit, with respect to linear coordinate transformations only. In doing so, they violated RLC's algorithm, if reluctantly:

> We must therefore leave open the question to what extent the general theory of the differential tensors associated with a gravitational field is connected with the problem of the gravitational equations. Such a connection would have to exist insofar as the gravitational equations are to permit [general coordinate transformations]; but in that case, it seems that it would be impossible to find *second-order* differential equations [that reproduce the Newtonian limit].[34]

It must have seemed to Einstein that at this point his hands were tied, so it is hard to fault him for his lack of faith in Ricci's method. Yet his conclusions would cost him three years of toil until he found his way back to the generally covariant equations of general relativity.[35]

***

In the 1900 review, RLC argue the merits of their calculus primarily as a facilitator for mathematical calculation and physical intuition:

> Poincaré has written that a good notation has the same philosophical importance in Mathematics as a good classification system has in the Natural Sciences. One may extend this remark, with even greater force, to cover methods. [...] The ADC is a natural tool for all research which deals with manifolds. [...] The generality and independence of choice of coordinates leads not only to elegance, but also to agility and insight into proofs and conclusions.[36]

This claim is borne out in the modern practice of general relativity, where the ability to choose coordinates arbitrarily is often exploited to perform calculations in the most expedient and physically transparent fashion.[37] Nevertheless, this significant advantage is overshadowed by the ADC's crucial role in conquering a *new* physical problem, one that RLC could not even have imagined in 1900: formulating a classical field theory (on the model of Maxwell's electromagnetism) *on a curved background*, while incorporating the key insight of the identity of gravitation and inertia.

In addition to its overwhelming success in explaining empirical observations, general relativity came to be recognized as a uniquely beautiful theory—thanks to the power and simplicity of its physical principles, but also to the formal elegance of its ADC-inspired equations. In Einstein's later years, dominated by his quest of a unified field theory that would span all physical interactions, Einstein elevated mathematical beauty to a guiding principle for theoretical physics:

> [Equations such as those of general relativity] can be found only through the discovery of logically simple mathematical conditions that determine the equations completely or almost completely. Once one has obtained [those], one requires only little knowledge of facts to set up a theory.[38]

Beyond such hyperbole, modern scholarship stresses how the actual development of general relativity was guided more by physical considerations (both calculations and thought experiments) than by a mathematical strategy informed by symmetry and covariance.[39] Furthermore, much of the beauty now perceived in general relativity is predicated on its profoundly geometrical character, very little of which is evident in Einstein's early treatments.[40]

A better understanding of the geometrical meaning of curvature had to wait for the work of Hessenberg,[41] Levi-Civita,[42] and Weyl.[43] For instance, the Riemann curvature tensor appears with this name only in Weyl's 1918 textbook, while the familiar interpretation of the covariant derivative as generating parallel displacement (i.e., propagating vectors and trajectories in the straightest way possible in curved space-time[44]) was introduced by Levi-Civita in 1917.

In the context of information technology, a *killer application* is a computer program so useful or desirable that it justifies the value of the hardware or operating system that hosts the application. It is not an exaggeration to state that Einstein's general relativity was the killer application for Ricci's ADC, propelling the latter from relative obscurity to acceptance as a fundamental field of mathematics, and as a foundational tool for modern physics.[45]

In contemplating the history of mathematics and physics through the lens of current research, it is tempting to focus on the heroic achievements of a few celebrated scholars, neglecting the role of their communities and of the larger currents of scientific thought that shaped and directed their work—in the case of Ricci and Einstein, the enduring discourse about the notion of invariance, examined along mathematical and physical avenues, respectively.[46]

Yet it remains subtly thrilling to encounter Ricci's legacy in everyday research (say, while calculating a covariant derivative, or computing the components of Ricci's tensor), and to let one's mind wander back to the rigorous and modest mathematician of Lugo, imagining the first time those expressions were written by his pen, and described by his carefully chosen words.

APPENDIX B

# T. Levi-Civita, "Gregorio Ricci-Curbastro"

Commemorazione del socio nazionale prof. Gregorio Ricci-Curbastro, letta dal socio Tullio Levi-Civita nella seduta del 3 gennaio 1926, has been translated into English by Donald Babbitt, David Goodstein, and Michele Vallisneri and is reprinted with permission of the publisher, the National Academy of the Lincei, from *Memoirs of the Academy of the Lincei,* 1 (1926), 555–564.

**Commemoration of national member Gregorio Ricci-Curbastro read by member T. Levi-Civita at the meeting of 3 January 1925[26].**

On January 8, 1881, Ricci began his inaugural course of mathematical physics at the University of Padua with some preliminary comments about the nature of the material to be covered. He then remarked,

> Students like you understand the immense advantages of these methods, which are especially useful for approaching what has been for a long time the ultimate goal of physics. I am speaking of the unity of science, which is the discovery of the relationships connecting various natural phenomena, and the causes on which they depend.[1]

Subsequently Ricci himself provided an essential contribution to this lofty goal, creating a new method—the absolute differential calculus—that gave Einstein the language to formulate mathematically his general theory of relativity.

This contribution alone would manifestly be enough to reveal the great merit of our Ricci, but allow me, who had the good fortune and honor to be his favorite disciple, as well as his colleague at Padua for more than twenty years—and tied to him even longer by constant and devoted friendship that only death could shatter—to recall our lamented colleague, whose unflinching moral rectitude, natural reserve, and the serenity with which he formulated judgments equaled the power of his intellect.

Ricci was born in Lugo on January 12, 1853, into an aristocratic and prominent Romagna family, the child of Antonio and Livia Vecchi. Together with his brother, Domenico, he was schooled privately in elementary studies and classics, acquiring a solid humanistic education and a solid understanding of the basic elements of mathematics and physics. In 1869 he easily passed the exam for admission to the Philosophical-Mathematical curriculum at the University of Rome.

Within the family, his decision to take up the study of mathematics was never questioned. While Ricci had shown himself to be an excellent student overall, he especially loved mathematics, even though his exceptional aptitude had not yet been recognized. At the same time his family considered it natural for him to select this branch of studies, which, at least in its applied aspects, followed naturally in the family tradition. In fact, his father was an expert engineer, and his mother was the daughter of Gregorio Vecchi. A student of Giuseppe Venturoli, he was the first professor of hydrometry in the Pontifical School of Engineering at Rome and afterward the chief engineer of the province of Bologna.

The quality of teaching and the level of scientific studies in Rome around 1870, were not, if truth be told, very high. Our Ricci, having completed the first course in mathematics, returned home and remained there for some time, following the wishes of his father, who was greatly disturbed by the events of September 1870. In 1872, Ricci enrolled in the University of Bologna and from there went on to Pisa, attracted by the fame of the Scuola Normale Superiore, where, following a competition, he entered as an external student in 1873.

His teachers there included Enrico Betti, Ulisse Dini, and Ernesto Padova, who later became a close friend.

In 1875 Ricci completed a carefully written dissertation on Fuchsian linear differential equations and obtained from this his *laurea* in physical and mathematical science with highest honors. The following year the Scuola Normale credentialed him to teach high-school mathematics, based on his second dissertation on a generalization of a problem posed by Riemann concerning hypergeometric functions. Like the first one, this dissertation contained original observations and deductions; neither thesis was published. His characteristic willingness to address mathematically rigorous questions in the most direct way was already evident in these youthful writings. But no less evident was his selfless love of science, which drove him to seriously study it, even if it did not advance his personal research. At some point, it also became apparent that that Ricci could no longer be content with just assimilating this broad scientific culture and needed to carry out original work. But that evolved gradually.

The recipient of a Lavagna scholarship in 1876-77, Ricci continued to attend Betti and Dini's lectures, which, in turn, paved the way for his first publications. Betti, who had come to appreciate the depth and mathematical acuity of his student's thinking, asked him to prepare several articles for *Nuovo Cimento,* summarizing the electromotive laws of galvanic circuits—which arise on the one hand from the elementary action formulas of F. Neumann, Riemann, and Clausius, and on the other from Maxwell's theory—while simultaneously noting several observations of a mechanical nature made by Betti himself in the course of that year. Ricci's articles provided a thoughtful and precise synthesis of these questions.

Another paper of Ricci's, published that same year in the *Giornale di Matematica,* shows Dini's influence. It considered a linear differential equation and its Lagrange adjoint and investigated the behavior of their respective solutions around a regular point, deducing, in particular, without any explicit calculations, that if the differential equation is reducible (over the complex analytic functions), then so was its adjoint.

In 1878–1879, supported by a fellowship that offered advanced study abroad, Ricci went to Munich, where he attentively followed the courses of Felix Klein and Alexander von Brill.

Klein, who energetically supported the ideas and research of the young mathematicians who had begun to flock to his school, took a liking to Ricci, who reciprocated with an outpouring of deferential and affectionate admiration. However, if Klein had any beneficial influence on the further development of his pupil's mathematical thought, the possible stimulus became so ingrained in Ricci's personality that no visible trace remained. This is in contrast to what has been observed in other of Klein's students, many of whom later became illustrious mathematicians themselves.

The following academic year, 1879–80, Ricci returned to Pisa as special assistant [*assistente straordinario*] to the chair of calculus, which Dini then held. After a public competition, Ricci was appointed associate professor [*professore straordinario*] of mathematical physics at the University of Padua, beginning December 1, 1880.

For Ricci, this professorial appointment marked the start of a period of immensely creative scientific activity. His first course gave him the opportunity to deepen the applicability and use of Green's functions in the theory of potentials. Subsequently he showed the equivalence of galvanic currents (constant in time) and permanent magnets. Starting with the mathematical expression of such equivalence, he proved it with great simplicity and generality using the explicit transformation of the respective potentials. He published these results in the *Atti del R. Ist. Veneto* in 1882 and the *Annali di Matematica,* 1884.

But his mind was already grappling with the fundamental problems of analysis and differential geometry. Through strenuous effort and beautiful preliminary investigations, he was led to the discovery and progressive refinement of the calculus that today bears his name.[2]

His first writings on this subject dealt with the construction of differential parameters, invariant differentials, and the classification of quadratic forms. Developing the ideas contained in these first attempts, Ricci gradually seized upon and perfected the calculations that emerged from them. He then modified the usual procedures employed in the differential calculus in such a way that the formulas and results always remained in the same form, no matter what system of coordinates was being used.

This explains why we have (and, as he showed it to be, a mathematically natural phenomenon) a system of functions[3] that behave in the same way when the coordinates are changed, independent of the choice of these coordinates. In addition, certain operations are introduced that are equally independent of the coordinates chosen, i.e. they are absolute, giving the name to the calculus.

The most useful applications arise when the nature of the material under consideration requires a quadratic differential form, i.e. as in differential geometry, where it expresses the infinitesimal distance between two points; in mechanics the kinetic energy; and in general relativity, the infinitesimal interval between two events in space-time. It is convenient then to take this differential form as given—that is, as absolute—and this is where the essential element of the new calculus arises: in the notion of covariant differentiation, which has the essential characteristics of ordinary differentiation, but also respects the invariant behavior (i.e., is independent of the choice of coordinates) of the system to which it is applied.

In 1869, Christoffel had, to be sure, pointed out such a differentiation, but it was Ricci who conceived and revealed it as an autonomous instrument. It permitted him to reduce to pure algebra, with unexpected insight and elegance, the problem of differential invariants and differential parameters belonging to a quadratic form and to others of arbitrary degree associated with it. In addition, Ricci was able to determine the conditions for the existence of orthogonal integrals for a partial differential equation of the first order. A particular application was to the theory of triple orthogonal systems of ordinary space.

Concerning this first phase of the development of Ricci's absolute differential calculus, it appears that it was appreciated only after the fact. With regard to his contributions, there exist two instructive commentaries prepared by [Eugenio] Beltrami[4] and [Luigi] Bianchi,[5] in 1887 and 1901 respectively, in connection with the competition for the Royal Prize in Mathematics.

In his commentary, Beltrami characterized Ricci's 1886 work as "important," concluding with this judgment: "It seems to us to represent the preliminary development of a powerful tool that appears to have already led to interesting results. However it still awaits its final justification by further tests in which perhaps its primitive and complex analytical apparatus could be replaced by simpler and more sophisticated approaches."

Thus Beltrami's commentary was an appreciative but cautious assessment of Ricci's work, recognizing the value of his findings, but reflecting his concerns about whether this new tool would lead to deep new results in the future. Indeed, while *a priori* confidence should never influence judgment, it never failed to spur Ricci's research. Over several years, he went on to show the usefulness of his methods. An example was his intrinsic characterization of surfaces for which $ds^2$ was reducible to Liouville's quadratic form. He also presented in a course his entire theory of surfaces in ordinary space, relating it to his absolute calculus of two variables in a most elegant and ingenious way. Unfortunately the lithographed notes for this course have been unavailable for many years.

Even more impressive than the application of his absolute differential calculus to surfaces was its application to objects involving three or more variables. Here the tool of choice for maximum simplicity and insight is the theory of orthogonal congruences of lines on a Riemannian manifold, a powerful generalization of the moving Triad in ordinary space. By introducing rotation coefficient and canonical congruences with respect to an assigned congruence, both ingeneous, profound innovations, Ricci fashions intrinsic geometry into a practical instrument for calculation. Thus in 1895, after about a decade of tireless work and successive conjectures, the absolute calculus reached its full and lush maturity.

Not long afterward, Ricci's attention was drawn to a question that, unknown to him, had just been posted for a competition by the Jablonowski Society of Leipzig: the question of classifying by way of curvature the three-dimensional spaces endowed with a group of symmetries. Already Bianchi had assigned the appropriate canonical forms to the problem, without, however, examining its invariant aspects. This made it the perfect subject for Ricci.

After rediscovering and illustrating the principal curvatures (which had been mentioned previously by Suvaroff and Schur), Ricci, taking as his starting point Killing's fundamental equations, suitably transformed, succeeded in establishing completely the requested criteria, which in the case of transitive groups assume a particularly simple and accessible form.

Armed with such useful mathematical tools, i.e., his absolute differential calculus, Ricci entered the contest for the Royal Prize in Mathematics for 1901. The prize was not conferred on him or anyone else that year, with the comment in his case that "the tools he developed are useful but not indispensable for treating various questions in mathematics." Ricci never complained, but kept the immutable conviction (which he held almost alone at the time) of having created a useful collection of mathematical tools.

Another fifteen years had to pass before the scientific world had proof of the validity of Ricci's work, through its essential role in Einstein's general theory of relativity, which also became for Ricci "the just dispenser of glories."

But let us return once again to his quiet existence in Padua, after he had obtained the chair of mathematical physics.

Quickly enveloped in the affectionate esteem of his colleagues, he became especially close to [Francesco] D'Arcais, who was a bit older than Ricci and who like him had been Dini's assistant. In 1882, when Ernesto Padova requested a transfer from Pisa to the chair of advanced mechanics at Padua, the old master and new colleague, who had common interests, rapidly became close friends. Although Ricci's teaching activity was intense during this period, and his attention was devoted mostly to those studies where the absolute calculus arose and flourished, he didn't just restrict himself to the contemplative circle of his fellow scientists, but participated actively in the administrative life of his hometown of Lugo as a provincial and municipal councilor. Among other things, he took an interest in practical hydraulics, which is attested to by two beautiful reports: one, prepared for the provincial council and filled with historic information, was entitled *On the hydraulic conditions of the countryside to the right of the Reno-Primaro and on the attempts to make it better*;[6] the other, also to the municipal council, was about a proposed aqueduct.[7]

Much later, the first project to which he contributed led to a culvert under the Santerno River. In 1924 when he and his sons visited the area he hinted to them, with justifiable pride, about his earlier written contribution to the project as well as his continued interest in its development during the years that followed.

In August of 1884 he married the lovely Bianca Bianchi Azzarani from Imola. Their marriage was blessed with great domestic happiness. They loved each other very much and led a quiet and tranquil life in a home embellished by her refined taste in art. Two sons and a daughter crowned this untroubled union, but it ended prematurely in 1914 when Bianca developed a malignant tumor that took her life within a short period of time.

* * *

As is well known, from the founding of the Kingdom of Italy to the end of the last century, professors and associate professors had distinct positions [meaning that there was no automatic path to promotion] in our universities. In particular, promotion to full professor [*professore ordinario*] depended on a very small number of positions being available, and as a result there were great disparities in how long it took an individual to go from associate professor to full professor. Very few managed it in the minimal required time of three years; the average wait was much longer, and there were occasions when a colleague who held on to his position for an especially long time blocked other worthy candidates from advancing.

Ricci's career was particularly affected by such a situation. He remained an associate professor for ten years, until December 10, 1890, when he was finally promoted, and then only because the minister [of public instruction] gave in to the repeated insistence of the faculty of science at the University of Padua and awarded him a newly created position of full professor, provided the faculty gave up the associate professor position that was becoming free because Giovanni Garbieri, who taught complementary algebra,[8] transferred to Genoa. Thus Ricci assumed the chair of algebra as a full professor, while continuing to teach his old course in mathematical physics as a lecturer.

The challenge of teaching two courses simultaneously was not unappealing to him, nor was the contact with a large number of students in his algebraic analysis class. In preparing lessons for them, he worked diligently to obtain the maximum conceptual clarity. In particular he gave an improved and rigorous exposition of Dedekind's theory of real numbers, relating this theory to and uniting it with the other major treatments of the real numbers, which may or may not adopt a rigid criterion in partitioning rational numbers in two classes. This course of algebra, which I attended for the first time, has forever remained for me a model of impeccable reasoning and fruitful mathematical enterprise, and not long after was printed in its entirety.

This first-year course was later changed profoundly as it evolved into one that sacrificed some elements of a traditional algebra course in favor of furnishing the essential elements of calculus. As a result it positioned students entering their second year to profitably attend not only a course in the most relevant parts of advanced pure and applied calculus but also an advanced course in rational mechanics. As the class continued to evolve, its lessons were collected and distributed as lithographed volumes, and during the last years of his life Ricci deemed the material ready for formal publication. That publication is in press at the publishing house of Milani in Padua,[9] with the author himself having done all the revisions to the proof sheets.

Even in this textbook, there is more than one original exposition, such as the systematic development of an enumerable representation in the theory of definite integrals, which had also been the subject of one of Ricci's articles.[10]

Ricci's lectures were not lively, but they were admirable for their precision and great fluidity of form: whoever wrote them down would never have had to edit them later on. Anyone who followed them attentively would have understood the core of the question being considered—which was always presented in great generality and took into account every possible subtlety—and would have sensed the vigor of Ricci's lucid intellect.

This clarity of thought, expressed with accessible eloquence and considerable dignity, meant that Ricci's advice was actively sought outside the fields of science. It was solicited not only in faculty councils—over which he presided from 1900 to 1908—and in the academic senate but also in public administration. From his youth in Romagna to his years in Padua, he was chosen as a candidate by the Catholics in administrative elections. Having become a city councilman as soon as an opportune situation (of party or coalition) presented itself, he was given executive responsibilities for public instruction, then finance. He held these offices with a zeal and equanimity that was recognized not only by his adversaries, but what is even rarer, by his own direct subordinates. Repeatedly asked to accept the position of mayor, he always declined because in his beautiful equipoise as thinker and citizen, he never allowed one to eclipse the other. This was fortunate for mathematics since he was continuing to produce new results.

* * *

The absolute calculus was completely developed by 1895. However in 1899, at the behest of Felix Klein, a summary exposition was written that was designed to be at the disposal of whoever needed it.[11] But for many years its use was almost exclusively limited to its inventor and a few of his students. After having reproduced with these methods Weingarten's theory of surfaces, Ricci used [his methods] to derive subtle metric properties of hyperspaces. We'll limit ourselves to recalling the double tensor derived from the contraction of two indices of the Riemann tensor[12] that yields the principal curvatures of a surface and their mean. Today this curvature is classical in relativity (in fact, it is Einstein's tensor $G_{ik}$[13] on four-dimensional space-time). No less beautiful from the mathematical point of view, if less rich in speculative potential, are his researches on three-dimensional manifolds that have intrinsic properties assigned *a priori*. We speak of simply stated, well-defined problems for which a formulation by ordinary methods presents itself with dismaying complexity: our Ricci managed with subtle analysis to classify and resolve them in exemplary style.

And, finally, when Ricci had passed his sixth decade, he received the most brilliant and incontrovertible affirmation of his absolute calculus.

After creating special relativity in 1905 and enjoying ardent acclaim and widespread, if not universal, agreement from his colleagues, Einstein himself was moved to modify what he had done, to express the laws of physics in invariant form with respect to reference frames in any state of motion, and more generally to any set of coordinates of space-time.

In a certain sense the question had already been resolved according to the classical scheme in which time stands alone [or even "does not mix with space as in relativity"] by the equations of Lagrange for mechanics and the contributions of Jacobi, Lamé, Beltrami, Padova, Hertz, and Volterra for other physical phenomena.

On the other hand, Ricci's methods, even for special relativity, would have easily accomplished the general transformation of coordinates (of space-time) assuming as the fundamental form $ds^2$ that expresses the infinitesimal "distance" in space-time.[14]

Mr. Kottler had shown this explicitly in 1912.[15]

But Einstein aimed to unify all physical phenomena, connecting geometry and gravitation. To accomplish this, it was necessary to give up an *a priori* given $ds^2$ and to substitute for it an *a priori* unknown four dimensional metric,[16] influenced by external circumstances (electromagnetic, optical, thermal, etc.), and interacting with them as both matter and background.

How can such an audacious goal be translated into precise formulas? In 1912, Marcel Grossmann, who was then a colleague of Einstein at the Swiss Polytechnic School in Zurich, had the remarkable insight that what Einstein needed was the absolute calculus. He brought it to Einstein's attention and began collaborating with him in a first attempt to utilize this new tool.[17] Einstein, having mastered this new calculus, didn't hesitate to use it to implement his extraordinary physical intuition, culminating in the construction of his celebrated gravitational equation, which turned out to be ideally suited to describing nature. Among other things it furnished the explanation of the displacement of the perihelion of Mercury, which had long been sought in vain using Newtonian mechanics.

The gravitational equations represent—these are Einstein's words—a real triumph of the methods of the absolute calculus created by Ricci.

Thus Beltrami's earlier commentary, alluded to above, expressing his concerns about whether or not Ricci's absolute calculus "would lead to new deep results in the future" was completely answered in the affirmative by Einstein himself.

It has been said that, like Apollonius of Perga, whose abstract investigations of the conic sections established the mathematical foundation for the astronomical and mechanical discoveries of Kepler, Galileo and Newton, Ricci did the same for general relativity. However the work of Apollonius lay fallow for eighteen centuries, while Ricci was more fortunate, in that he was able to witness the superb way in which his theories were put to work.

Profoundly admired by a small group of his colleagues, he remained unknown to many others because of his modesty and reserved character, which largely kept him from engaging in scientific interactions that accompany modern life, such as frequent travels, personal contacts, contributions at congresses and conferences, and the relatively wide distribution of his publications, both original and expository.

Official recognition of Ricci's great contributions was late in coming, on the part of both academia and government. He was recognized by the Venetian Institute in 1892 (and he became its president for the biennium 1916–1918). But only in recent years did the Academy of Turin (1918), the Society of the XL (1921), the Royal Academy of Bologna (1922), and the Pontifical Academy (1925) elect him a member. In 1905, the Academy of Padua named him a corresponding member, and in 1915 a permanent member. Our academy, which was pleased to have published the first communication of his principal results, named him a correspondent in 1899 and a national member in 1916.

The growing public awareness of general relativity assured that his methods were, in particular, brought to the attention of people educated in mathematics and physics. As a result this work inspired countless books dealing with relativity and firmly established the luminous position of Ricci's work in the history of science.

Padua and Lugo, the two cities in which he carried out his invaluable work as an eminent mathematician, an exemplary teacher, and wise administrator, have hastened to honor his memory with memorial tablets, busts and other monumental statuary, which are tangible testaments to his academic achievements and his exemplary citizenship.

* * *

He passed away on August 6, 1925, in a hospital in Bologna. He had entered it on July 21, having completed his teaching duties and enduring the prolonged stresses of the exam period, in which he participated against the explicit advice of his doctor. In the hospital he had an operation to relieve him of a painful bladder disorder that had tormented him for years, but which was not thought to be of an alarming nature. The surgical procedure was seemingly successful: He was out of bed and looking forward to a recovery that would have allowed him to dedicate himself with renewed vigor to his research. But his heart wouldn't cooperate, and in a matter of hours a sudden attack of angina pectoris shattered his robust constitution.

Without ever drawing attention to it, he amply practiced charity. This was in keeping with his last will and testament, in which, with moving tenderness, he encouraged his sons to be lavish givers. He requested a funeral without pomp, arranging that the family tomb in the cemetery of Lugo should display in his memory a simple tablet professing his profound devotion to his Catholic faith and summing up his entire life in this pronouncement:

"He was for 45 years professor of mathematics at the University of Padua."

This was an edifying example of modesty in a man who knew full well that his name would be forever associated with the absolute differential calculus and its magnificent applications!

APPENDIX C

# Obituary of Tullio Levi-Civita

TULLIO LEVI-CIVITA

1873—1941

Tullio Levi-Civita

[Reprinted without change of pagination from *Obituary Notices of Fellows of The Royal Society*, Volume 4, November 1942]

# TULLIO LEVI-CIVITA

## 1873–1941

THE death of Tullio Levi-Civita, following within fifteen months on that of Vita Volterra, removes from the roll of foreign members of the Royal Society the last representative of a great school of mathematics. Both of these mathematicians had in the course of active lives contributed greatly to the high reputation enjoyed by Italian mathematics in general, and the school of mathematics in Rome in particular; both had made many contributions which have found a permanent place in mathematical literature, and both ended their days as victims of a political system which destroyed institutions and liberties in which they were firm believers.

Levi-Civita was born in Padua on 29 March 1873, the son of Giaccomo Levi-Civita and his wife, Bice Lattis. The family was a wealthy one, well known for its strong liberal traditions. Giaccomo Levi-Civita was a barrister, jurist and politician, and was for many years mayor of Padua, and a Senator of the Kingdom of Italy. As a young man he had served as a volunteer and fought with Garibaldi in the campaign of 1866, and he had played an important part in the Risorgimento.

Giaccomo Levi-Civita was anxious that his son should follow in his footsteps as a barrister, but Tullio's interest in the physical and mathematical sciences was apparent even in early childhood, and when he expressed a wish to follow his own inclinations his father never opposed him; and in later years the son's eminence in the scientific world was a source of great pride to the father. Consequently, when he completed his classical studies at the Ginnasio-Liceo Tito Livio in his native city at the age of seventeen, Tullio Levi-Civita entered the faculty of science at the university of Padua as a student of mathematics, and four years later he took his degree.

Amonst his teachers at the university of Padua were D'Arcais, Padova, Veronese, and Ricci-Curbastro (known to the scientific world simply as Ricci). The two last-named were the most distinguished, and both had considerable influence on the future career of their brilliant pupil. The influence of Ricci is the more obvious, since it developed into active collaboration, but probably Veronese's influence was quite as important, since it is largely to him that Levi-Civita owed the remarkable spacial intuition and familiarity with multidimensional space which characterizes the younger man's contributions to the Ricci-Levi-Civita partnership in the absolute differential calculus.

Levi-Civita's undergraduate days were not over before he began to write mathematical papers, and his ability was quickly recognized. Indeed, in the year after he took his degree, his application for the chair of mechanics at Massina was strongly favoured by Volterra and Morera, two of the electors; but the other two electors and Cremona, the chairman, supported Marcolongo, who was considerably Levi-Civita's senior both in years and in experience, and he was consequently elected. Three years later Padova, one of his former teachers at Padua, died, and Levi-Civita was elected in his place to the chair of mechanics, at the age of twenty-five. For twenty years he held this post, and these were among the most productive of his life. In 1918 he was called to the chair of mechanics at Rome, a post which he held for another twenty years, until racial discrimination, introduced into Italy in 1938, brought about his removal from office. Until then his life was uneventful, spent in the happy pursuit of his mathematical interests.

Levi-Civita's researches covered a vast field of mathematics, and it is not possible to say he was this or that kind of mathematician; one can only say he was a mathematician, and a great one. Viewing his work as a whole, however, the dominating impression one receives is of an astounding command of the technicalities of pure mathematics, aided by an acute geometrical intuition, applied mainly to problems of applied mathematics. This is, of course, only a general impression, for there is plenty of Levi-Civita's work which is first-class pure mathematics, and plenty that is genuinely applied. But one sometimes feels that parts of his work are insufficiently appreciated simply because on the one hand so much of it deals with special problems outside the range of interest of the pure mathematician, and on the other hand it is regarded as too theoretical by the applied mathematician. This fate has befallen other mathematicians who can be described as pure mathematicians whose main interest is in applying their knowledge to physical problems, but in Levi-Civita's case some of his contributions in this no-man's land between pure and applied mathematics have been of such importance that they could not be hidden; and in particular his contributions to the absolute differential calculus have placed both pure and applied mathematicians under a debt of gratitude which they have been glad to acknowledge.

But however much Levi-Civita's interest may have moved towards applied mathematics, his very earliest paper was quite unambiguously on pure mathematics. While he was a student at Padua his teacher, Veronese, had started a discussion on non-Archimedean geometry, and in 1893, before taking his degree, he contributed a paper to the discussion which was published in the *Atti Lincei*. After this one paper Levi-Civita does not seem to have taken any further part in the discussion, though Veronese became involved in a heated argument, mainly with Peano, until five years later when he published a couple of notes in the same journal which effectively resolved the difficulties and finally answered the question under discussion.

Some of Levi-Civita's other early papers suggest that his first attraction was towards pure mathematics. Thus in 1895 he published an improvement on Riemann's formula for expressing the number of prime numbers lying in a given interval as a contour integral. To about this same time also belongs a paper devoted to the solution of the integral equation

$$u(x) = \int_{a(x)}^{(bx)} f(x, y)\, v(y)\, dy$$

for the function $v(y)$, and this was followed by a paper applying the result to a problem in electrostatic induction. This is apparently the first of the many occasions on which he used his skill as a pure mathematician to solve a problem of physical importance. Other such occasions now followed in rapid succession.

To this same period, the closing years of last century, belong Levi-Civita's first researches in differential geometry and the absolute differential calculus, and broadly speaking we may say that by the time Levi-Civita was twenty-five he had established himself as a mathematician and had settled down to a continuous programme of work of the type he found most congenial to him. From then on papers appeared continuously in which he displayed great analytical skill in the solutions of all sorts of problems in applied mathematics. One can best review his work by abandoning the chronological order and considering his work as a whole in the various fields to which he contributed.

The work by which Levi-Civita is certainly best known is that on the absolute differential calculus, with its applications to relativity theory. The study of the particular class of invariants known as tensors goes back to the work of Riemann and Christoffel on quadratic differential forms (though the name tensor was only introduced by Voigt in 1898). In 1887 Ricci published his famous paper in which he developed the calculus of tensors including the important operation of covariant differentiation. For a considerable number of years following the publication of this paper he was engaged in working out his 'absolute differential calculus' aided by a number of able pupils, foremost among them being Levi-Civita. The results of the work of Ricci, Levi-Civita and others were finally published in a joint memoir by Ricci and Levi-Civita which appeared in 1900 under the title, 'Méthodes de calcul differential absolu et leurs applications', and which presented the theory of tensors essentially in the form used by Einstein and others fifteen years later.

In 1917 Levi-Civita made an advance in the absolute differential calculus of fundamental importance, with the introduction of the concept of parallel displacement. In dealing with parallelism in a plane one is accustomed to think of a vector at $P$ parallel to a vector at $Q$ simply as a vector at $P$ with a particular property, not as one derived from the vector at $Q$ by a certain operation. By considering the process of passing from one vector to another, Levi-Civita was led to his conception of parallelism on any surface. Parallelism on a developable surface can obviously be defined in terms of parallelism on the plane to which the developable is applicable. From these considerations Levi-Civita was led to define, on any surface whatsoever, the vector at $P$ parallel to a given vector at $Q$, as follows. He selected a path $C$ joining $P$ to $Q$ and considered the developable formed by the tangent planes to the given surface along $C$. The vector at $P$ parallel on this developable to the given vector at $Q$ is defined as that obtained by parallel displacement along *C*. It depends, of course, on the choice of the path C. It is easy to write down the differential equations for the parallel displacement along a curve of a vector on a surface and the form of these equations suggests immediately the law of parallel displacement in any Riemannian space.

Few mathematical ideas have found such diverse applications so quickly as that of parallel displacement. It is the basis of the unified representation of gravitational and electromagnetic fields in relativity theory, and there are still more far-reaching consequences which are not yet fully recognized in physics. The idea is no less important in pure mathematics. About the same time as Levi-Civita published his great paper on parallel displacement, Hessenberg published a paper in which it was shown that the notion of covariant differentiation did not depend in any essential way on a Riemannian metric, but was capable of considerable generalization. The ideas of the two papers were quickly taken up and developed, first by Weyl who used them for his unified theory of gravitation and electromagnetism, then by many others, and from them has developed the whole of the modern differential theory of generalized spaces.

Levi-Civita's direct contributions to relativity theory are substantial, but they are of a less conspicuous nature. From 1917 to 1919, in a series of papers, he and his students treated very elegantly the problems arising in the special case of a static gravitational field, including systems in a state of steady rotation. In 1937 he announced a result which, although it has proved to be erroneous, drew attention to a difficult and interesting question and greatly stimulated the development of relativistic mechanics. Since no exact solution of the problem of two bodies in relativity theory has been found, much attention has been paid to methods of approximation, which often involve very laborious methods of calculation as well as subtle points of theory. In pursuing this problem, Levi-Civita reached the conclusion that (relatively to distant objects) the centre of gravity of a double star has a secular acceleration in the direction of the major axis of the orbit, towards the periastron of the larger mass. This result could not be said to be inconsistent with recognized principles, but it was sufficiently surprising to awake keen suspicion. Direct criticism was scarcely possible, the calculation being too extensive for detailed publication. The obscurity deepened when it was found that de Sitter's earlier formulae led to a similar result but disagreed as to the magnitude. The problem was re-investigated by Einstein, Infield and Robertson, and by Eddington and Clark: these two investigations agreed in contradicting the supposed acceleration. Ultimately Levi-Civita found an algebraic mistake in his own calculation and came into agreement. Though in one sense a failure, his intervention in the two-body problem greatly benefited relativity mechanics which had been languishing for want of a definite aim.

In addition to the large number of papers which Levi-Civita published on the absolute differential calculus and relativity he published two books, *Questioni di meccanica classica e relativistica* (1924) and *Lezioni di differentiale assoluto* (1925). Both of these have become standard works, and the latter was translated into English in 1927 by Miss Marjorie Long and has been widely read in the translation, as well as in the original.

While he is most famous for his work in relativity theory, Levi-Civita has had an important influence on many other branches of mathematics. It is particularly necessary to mention his work in analytical dynamics, to advance which subject he did as much as any one during the earlier years of the twentieth century. Most of this work is however too detailed to permit of any analysis in this notice, and it must be sufficient if we mention some of the more important. The largest individual group of these papers deals with the problem of three bodies, either in the general case or in the restricted case in which two of the bodies describe circular orbits while the mass of the third is so small that it does not affect the motion of the other two particles. In these papers Levi-Civita has done much to increase our knowledge of the types of analytic solution possible, and he has also found an analytic condition for a collision resulting from given initial conditions, in the restricted case.

There is also a group of papers dealing with various problems that arise in connexion with the integrals of the equations of a general dynamical system. Perhaps the most important of these is one in which he establishes the existence of a system of integrals of a Hamiltonian system corresponding to given invariant relations. Other papers on dynamics deal with such a diversity of subjects as the analytic solution of Kepler's equation connecting the mean anomaly with the eccentric anomaly and eccentricity of a particle moving in an elliptic orbit, the effect of neglecting terms in the criteria for the stability of orbits, the Kowalevsky top, and Saturn's rings. And here we must also mention Levi-Civita's important contributions to the theory of adiabatic invariants. In 1923 Levi-Civita, in association with Professor Amaldi, published a three volume work on rational mechanics, *Lezioni di meccanica razionale*, which is now one of the accepted classics on the subject, and which has been translated into various other languages, including Russian.

Hydrodynamics is another subject which attracted Levi-Civita's attention, and to which he made a considerable number of contributions. Most of these deal with the solution of problems in classical hydrodynamics which require considerable analytical skill, but it is perhaps true to say that his work is too theoretical to appeal to modern experts in hydrodynamics. Nevertheless he has performed an extremely useful function by supplying much needed rigour at several points of the theory. His work on hydrodynamics is to a certain extent bound up with his work on the general theory of systems of partial differential equations. His work on this subject forms an important addition to the well-known Cauchy-Kowalevsky theory. An excellent account of this is given in a booklet, *Caratteristiche dei sistemi differenziali e propagazione ondosa*, which he published in 1931 and which was later translated into French. The Cauchy-Kowalevsky theorem affirms the existence of regular solutions of a normal system of partial differential equations in the neighbourhood of given initial values. The equations of the small motion of a fluid or of an elastic solid, the equations of the electromagnetic field, as well as Schrödinger's equation, all come under this theory. Levi-Civita's work is largely concerned with solutions having given initial values in a neighbourhood in which the equations cease to be normal, and his booklet is concerned not only with the general theory but with special cases of physical interest; thus, for example, he is able to deduce the impossibility of the propagation of a discontinuity in a viscous medium. He was particularly interested in the relation between the classical theory of wave propagation and modern wave mechanics, a topic to which he devoted a considerable part of an address which he gave in Chicago to a joint meeting of the American Mathematical Society and the American Association for the Advancement of Science in 1933.

It was natural that the efforts made to find a common framework to contain both quantum mechanics and the general theory of relativity should prove of the greatest interest to Levi-Civita. In 1933 he published a paper in which he proposed to replace Dirac's first order equations by a set of second order equations which took into account the gravitational field. When the two sets of equations are compared in the case in which there is no gravitational field it is found that Levi-Civita's equations are reconcilable with Dirac's when the electromagnetic field is either purely electric or purely magnetic, but not in the general case.

Though reference has been made to some of Levi-Civita's more outstanding researches there are very many which must be passed over. It ought to be mentioned, however, that in addition to his work in the realm of pure science he was frequently consulted by technicians and engineering firms on problems of practical engineering. In this way he was brought to do work of considerable value to the outside world, notably in connexion with the construction of submarine cables and the vibration of bridges.

Again, there is a great deal of evidence of Levi-Civita's interest in branches of mathematics in which he never published any researches. Old pupils frequently consulted him on their problems, and always found him acquainted with their work, even though it differed widely from anything that Levi-Civita had taught them, and he was always ready both with encouragement and with useful advice. He never took any part in the activities of the school of algebraic geometry headed by his colleagues Castelnuovo, Enriques and Severi in the university of Rome, but he was well informed on the work of that school. I remember an occasion on which I had written a paper on Abelian integrals which had provoked some discussion; I received a request to send reprints to certain mathematicians in Rome, and the names of those asking were Castelnuovo, Enriques, Severi and Levi-Civita.

Let us now try to form some idea of the background to all this scientific activity. For most of his life Levi-Civita was very fortunate in his private life, and being of a very friendly disposition he had a very happy life, free from any crisis having a bearing on his work, until tragedy came in his latest years.

Levi-Civita was not at first appearances a very striking figure. The first impression received was of an exceedingly small man (he was only about five feet tall) who was very short-sighted. His bearing was quite unpretentious; but, after having talked to him for a while, one was particularly struck by the vivaciousness and precision of his discourse and his very wide knowledge extending over pure and applied mathematics, astronomy and physics, and also by his remarkable acquaintance with the literature on these subjects, both old and new. One noticed in particular his quick grip on a problem, and his passionate interest in all sorts of scientific questions.

In spite of his small frame Levi-Civita was very robust, and enjoyed excellent health until he was well over sixty. What energy he had to spare from his work was devoted to his three great hobbies—mountaineering, cycling, and foreign travel. As a young man he devoted most of his vacations to mountaineering in the Dolomites, and in spite of his physical handicaps he was a good climber. As he grew older his worsening eye-sight curtailed his climbing activities, but he kept them up as far as he could for many years. He enjoyed cycling, and was often to be seen cycling round the countryside near Padua while he was professor there, and subsequently during his frequent visits to his parents; and when he was no longer able to climb he continued to visit his beloved mountains on his bicycle.

He was singularly well placed for indulging his third passion, foreign travel. His private fortune and freedom from domestic worries removed many obstacles from his way, and opportunities were regularly given him by invitations to visit countries in all parts of the world in order to deliver lectures to scientific gatherings. These visits he enjoyed intensely; he could see new places, meet new people, and, thanks to his own personal charm, make a host of new friends. Indeed, Levi-Civita was one of the personally best known and best liked mathematicians of his time.

Levi-Civita was also fortunate in his home life. There was a strong bond of affection between his father, who died in 1922, and himself. While the father was intensely proud of his son's scientific achievements, the son was equally proud of the father's record in the Italian wars of liberation. Though not himself an active politician, Levi-Civita was extremely interested in politics, and remained true to the noble traditions of his family. A visitor to his study could be in no doubt as to his beliefs; while three of the walls were lined by bookcases, the fourth remained empty save for two solitary portraits, one of his father and the other of Garibaldi. Levi-Civita always felt a great tenderness for his mother, and visited her regularly, either at the family house in Padua, or at her villa in Vigodarzere, a nearby village where she lived for several years before her death in 1927. After her death the villa was kept on by her daughter, Ida Senigaglia, and Levi-Civita continued to spend a part of each year there.

In 1914 Levi-Civita married Libera Trevisani. She had been his pupil at the university of Padua, and had taken her doctorate in mathematics. She proved herself a clever and affectionate companion to him, and she was a very gracious hostess, not only to the many eminent scientists who came to visit them from afar, but to the more humble students who regularly visited their home. She accompanied her husband on his many travels and shared with him the many friendships which he made on these journeys. There were no children of the marriage.

For Levi-Civita, research and teaching went hand in hand, and he guided a great number of pupils in fields in which he was the pioneer. His teaching was not circumscribed by any curriculum, as was usual in an Italian university, but was freely given to all who came to consult him. During his vacations, either in the Dolomites or at Vigodarzere, his former pupils would come to be near him, and he would follow their researches with the utmost interest. With infinite patience and unselfishness he would enter into the problems which they brought to him; nothing gave him more pleasure than to have an opportunity of helping them, and it was with the greatest pride that he would present their works for publication by one or other of the many learned societies of which he was a member. Indeed, as one of his pupils once remarked, no one ever merited more than he did the title of Maestro.

He was a born teacher. His scientific papers are models of lucidity, and his books are amongst the easiest reading on their specialized topics. In conversation he could give in simple terms a very simple account of an abstract theory. In acquiring his command over such vast fields of science he was greatly aided by being the possessor of an unusually good memory.

Many honours came to Levi-Civita. He received honorary degrees from many universities throughout the world, including Toulouse, Aachen, Amsterdam and Paris. Academies in all countries of the world honoured him by election to honorary membership. The list is too long to repeat, but included the Institut de Paris and the Berlin academy, and societies in Leningrad, Madrid and South America. In his own country he was a member of the Reale Accademia dei Lincei, the Reale Accademia d'Italia and the Pontificia Accademia delle Scienze dei Nuovi Lincei. When, in 1936, Pope Pius XI dissolved the last named and replaced it by the Pontificia Accademia delle Scienze, an international body, Levi-Civita became an original member of the new academy. The Sylvester medal of the Royal Society was conferred on him in 1922, and in 1930 he was elected a foreign member. He was elected an honorary fellow of the Royal Society of Edinburgh in 1923, and of the London Mathematical Society in 1924. He was also an honorary member of the Edinburgh Mathematical Society, and attended one of its colloquia in St Andrews.

Although he did not take any active part in politics, he could not remain indifferent to the rise of fascism in Italy, and in 1925, after the 'Matteotti affair' he was, with Volterra and many other Italian scientists, a signatory of the 'Manifesto Croce', which stigmatized fascism and deprecated its growing power in Italy. For some time, however, his scientific renown protected him from persecution. But in September 1938 the government issued decrees removing from office all professors of Italian universities who were of Jewish origin, and dismissing them from Italian academies. Levi-Civita, who came under the ban, found himself cut off from all that made life interesting to him. It is recorded that he learned of the decrees while staying with his sister at Vigodarzere, when some one happened to switch on the radio as they were being announced. His expression did not change, and he went out for his afternoon walk as usual. The blow soon told on him, however. His health began to fail. After he returned to Rome severe heart trouble developed, and as he was forbidden by his doctors to take any long journey he could not accept any of the offers of asylum which came to him from foreign universities. The remaining years of his life were very sad; he was confined to his room, and unable to continue his work. He died on 29 December 1941 of a stroke. At first the Roman newspapers, except the *Osservatore Romano*, the organ of the Vatican, ignored his death, and it was only after the Pontifical Academy had used its influence that the family were able to announce in the newspapers the fact of his death and the arrangements for his funeral.

In preparing this notice I have received valuable assistance from Dr Enrico Volterra, Professor Beniamino Segre, Sir Arthur Eddington, Mr L. A. Pars and Professor E. T. Whittaker.

W. V. D. HODGE

## BIBLIOGRAPHY

1893. Sugli infiniti ed infinitesimi attuali quali elementi analitici. *Atti Ist. Veneto*, **51**, 1775–1815.

1894. Sugli invarianti assoluti. *Atti Ist. Veneto*, **52**, 1447–1523.

1895. Sui gruppi di operazioni funzionali. *R. C. Ist. Lombardo*, (2) **28**, 458–468.

1895. Alcune osservazioni alla nota—Sui gruppi di operazioni funzionali. *R. C. Ist. Lombardo*, (2) **28**, 864–873.

1895. I gruppi di operazioni funzionali e l'inverzione degli integrali definiti. *R. C. Ist. Lombardo*, (2) **28**, 529–544, 565–577.

1895. Di una espressione analitica atta a rappresentare il numero dei numeri primi, compresi in un determinato intervallo. *R. C. Accad. Lincei*, (5) **4**, (i) 303–309.

1895. Sull' inversione degli integrali definiti nel campo reale. *Atti Accad. Torino*, **31**, 25–51.

1895/1896. Sulla distribuzione indotta in un cilindro indefinito da un sistema simmetrico di masse. *R. C. Accad. Lincei*, (5) **4**, (ii) 332–336; (5) **5**, (i) 34–40.

1896. Sugli integrali algebrici delle equazioni dinamiche. *Atti. Accad. Torino*, **31**, 816–823.

1896. Sulle transformazioni delle equazioni dinamiche. *Ann. Mat. pura appl.* (2) **24**, 255–300.

1896. Sul moto di un corpo rigido intorno ad un punto fisso. *R. C. Accad. Lincei*, (5) **5**, (ii) 3–9, 122–127.

1896. Sul moto dei sistemi con tre gradi di libertà. *R. C. Accad. Lincei*, (5) **5**, (ii) 164–171.

1896. Sul moto di un sistema di punti materiali, soggetti a resistenze proporzionali alle rispettive velocità. *Atti Ist. Veneto*, **54**, 1004–1008.

1897. Sur les integrales quadratiques des équatians de la mécanique. *C. R. Acad. Sci. Paris*, **124**, 392–395.

1897. Sur une classe de $ds^2$ à trois variables. *C. R. Acad. Sci. Paris*, **124**, 1434–1438.

1897. Sulla riducibilità delle equazioni elettrodinamiche di Helmholtz alla forma hertziana. *Nuovo Cim.* (4) **6**, 93–108.

1897. Sopra una classe di integrali dell' equazione $A^2\ d^2V/dt^2 = d^2V/dx^2 + d^2V/dy^2$. *Nuovo Cim.* (4) **6**, 204–209.

1897. Sulla stabilità dell' equilibrio per i sistemi a legami completi. *Atti Ist. Veneto*, **55**, 1247–1250.

1898. Sui numeri transfiniti. *R. C. Accad. Lincei*, (5) **7**, (i) 91–96, 113–121.
1898. Sulla integrazione dell' equazione $\triangle^2 \triangle^2 u=0$. *Atti Accad. Torino*, **33**, 932–956.
1898. Sopra una trasformazione in sè stessa della equazione $\triangle^2 \triangle^2 u=0$. *Atti Ist. Veneto*, **57**, 1399–1410.
1899. Sulle congruenze di curve. *R. C. Accad. Lincei*, (5) **8**, (i) 239–246.
1899. Sulle equazioni a coppie di integrali ortogonali. *R. C. Accad. Lincei*, (5) **8**, (i) 295–296.
1899. Tipi di potenziali, che si possono far dipendere da due sole coordinate. *Mem. Accad. Torino*, (2) **49**, 105–152.
1899. Sur les intégrales périodiques des équations aux dérivés partielles du premier ordre. *C. R. Acad. Sci. Paris*, **128**, 978–981.
1899. Interpretazione gruppale degli integrali di un sistema canonico. *R. C. Accad. Lincei*, (5) **8**, (ii) 235–238.
1900. Complementi al teorema di Malus-Dupin. *R. C. Accad. Lincei*, (5) **9**, (i) 185–189, 237–246.
1900. Sur l'instabilité de certaines substitutions. *C. R. Acad. Sci. Paris*, **130**, 103–106.
1900. Sur l'instabilité de certaines solutions périodiques. *C. R. Acad. Sci. Paris*, **130**, 170–173.
1900. Sur le problème restreint des trois corps. *C. R. Acad. Sci. Paris*, **130**, 236–239.
1900. Funzioni armoniche e trasformazioni di contatto. *Atti Ist. Veneto*, **59**, 671–675.
1900. (With G. Ricci). Méthodes de calcul différentiel absolu et leur applications. *Math. Ann. B*, **54**, 125–201. [Polish translation, Warsaw 1901.]
1901. Sopra alcuni criteri di instabilità. *Ann. Mat. pura appl.* (3) **5**, 221–308.
1901. Sulla determinazione di soluzioni particolari di un sistema canonico quando se ne conosce qualche integrale o relazione invariante. *R. C. Accad. Lincei*, (5) **10**, (i) 3–9, 35–41.
1901. Sui moti stazionari dei sistemi olonomi. *R. C. Accad. Lincei*, (5) **10**, (i) 137–143.
1901. Sui moti stazionari di un corpo rigido nel caso della Kowalevsky. *R. C. Accad. Lincei*, (5) **10**, (i) 338–346, 429–434, 461–466.
1901. Sulla resistenza dei mezzi fluidi. *R. C. Accad. Lincei*, (5) **10**, (ii) 3–9.
1901. Sulla forma dello sviluppo della funzione perturbatrice. *Atti Ist. Veneto*, **60**, 653–661.
1901. Sui massimo cimento dinamico nei sistemi elastici, *Nuovo Cim.* (5) **2**, 188–196.
1902. Sur le champ électromagnétique engendré par la translation uniforme d'une charge électrique parallèlement à un plan conducteur indéfini. *Ann. Fac. Sci. Toulouse*, (2) **4**, 5–44. [Also in Italian, 1903, *Nuovo Cim.* (5) **6**, 141–185.]
1902. Influenza di uno schermo conduttore sul campo elettromagnetico di una corrente alternativa parallela allo schermo. *R. C. Accad. Lincei*, (5) **11**, (i) 163–170, 191–198, 228–237; also *Nuovo. Cim.* (5) **3**, 442–455.
1902. La teoria elettrodinamica di Hertz di fronte ai fenomeni di induzione. *R. C. Accad. Lincei*, (5) **11**, (ii) 76–81.
1902. Sulla cinetostatica. *Atti Accad. Padova*, **18**, 145–150.
1902. Sur les surfaces (*S*) di M. Zaremba. *Bull. Acad. Sci. Cracovie*, 1902, 264–270.
1902. Sur les fonctions de genre infini. *Bull. Sci. Math.* (2) **25**, 333–335.
1903. Sur les trajectoires singulières du problème restreint des trois corps. *C. R. Acad. Sci. Paris*, **135**, 82–84.
1903. Condition du choc dans le problème restreint des trois corps. *C. R. Acad. Sci. Paris*, **135**, 221–223.
1903. Traiettorie singolari ed urti nel problema ristretto dei tre corpi. *Ann. Mat. pura appl.* (3) **9**, 1–32.
1903. Sur la singularité dont sont affectées, pour une vitesse nulle, les équations du mouvement d'un point frottant sur une surface. *Arch. Math. Phys.* (3*B*) **5**, 28-37.

1904. Sopra la equazione di Kepler. *R. C. Accad. Lincei*, (5) **13**, (i) 260–268; also *Astr. Nachr.* no. 3956, 313–314.
1904. Sopra un problema di elettrostatica che interessa la construzione dei cavi. *R. C. Accad. Lincei*, (5) **13**, (i) 375–382; also *Nuovo Cim.* (5) **8**, 187–195.
1904. Sulla integrazione della equazione di Hamilton-Jacobi per separazione di variabili. *Math. Ann. B*, **59**, 383–397.
1905. Sopra un problema di elettrostatica che si è presentato nella construzione dei cavi. *R. C. Circ. Mat. Palermo*, **20**, 173–228.
1905. Sulla ricerca di soluzioni particolari dei sistemi differenziali. *R. C. Accad. Lincei*, (5) **14**, (i) 203–209.
1905. Sulle funzioni di due o più variabili complesse. *R. C. Accad. Lincei*, (5) **14**, (ii) 492–499.
1906. Sur la résolution qualitative du problème restreint des trois corps. *Acta Math. Stockh.* **30**, 305–327. [Summary in *Verh.* 3 *Int. Math. Congr.* (Leipzig, 1905), 402–408.]
1906. Sulla contrazione delle vene liquide. *Atti Ist. Veneto*, **64**, 1465–1472.
1906. Sur la recherche des solutions particulières des systèmes différentiels et sur les mouvements stationnaires. *Math.-phys. Abh. Warschau*, **17**, 1–40.
1906. The mixed transformations of Lagrange's equations. *Nature, Lond.* **74**, 516.
1906. Sulla penetrazione dei proiettili nei mezzi solidi. *Atti Ist. Veneto*, **65**, 1149–1158.
1906. Ueber einer technische Aufgabe die in Beziehung zur konformen Abbildung steht. *Verh. Ges. dtsch. Naturf. Aerzt.* **77**, 20–21.
1907. Le idee di Enriques sui principi della meccanica. *Riv. Filos. Sci. Aff.* **9**, 337–346.
1907. Sullo sviluppo delle funzioni implicite. *R. C. Accad. Lincei.* (5) **16**, (ii) 3–12.
1907. Sur le mouvement de l'électricité sans liaisons ni forces extérieures. *C. R. Acad. Sci. Paris*, **145**, 417–420.
1907. Sulla massa elettromagnetica [report] *Nuovo Cim.* (5) **14**, 271–304; also *Riv. Sci.* **2**, 387–412.
1907. Sulle onde progressive di tipo permanente. *R. C. Accad. Lincei*, (5) **16**, (ii) 776–790.
1907. Giuseppe Picciati [obituary]. *Nuovo Cim.* (5) **15**, 363–368.
1908. Sui campi elettromagnetici puri dovuti a molti piani permanenti. *Atti Ist. Veneto*, **67**, 995–1010.
1908. Sull' attrazione esercitata da una linea materiale in punti prossimi alla linea stessa. *R. C. Accad. Lincei*, (5) **17**, (ii) 3–15.
1908. Sull' attrazione newtoniana di un tubo sottile. *R. C. Accad. Lincei, R. C. Accad. Lincei*, (5) **17**, (ii) 413–426, 535–551.
1909. Sulle azioni meccaniche dovute ad un flusso filiforme di elettricità. *R. C. Accad. Lincei*, (5) **18**, (i) 41–50.
1909. Teoria asintotica dei potenziali elettriche. *R. C. Accad. Lincei*, (5) **18**, (i) 83–93.
1909. Sulla espressione asintotica dei potenziali ritardati. *Atti* 4 *Congr. Int. Mat.* (Roma), **3**, 93–100.
1909. Sulla forma dell' anello di Saturno. *Atti Ist. Veneto*, **68**, 557–583.
1909. Sulla costituzione delle radiazioni elettriche. *Nuovo Cim.* (5) **18**, 163–169.
1909. Commemorazione del Socio Valentino Cerruti. *R. C. Accad. Lincei*, (5) **18**, (ii) 565–575.
1910. Sul teorema di esistenza della funzioni implicite. *Atti Ist. Veneto*, **69**, 291–302.
1910. Ueber Lorentz-Einstein'sche starre Bewegungen. *Ann. Phys. Lpz.* (4*B*) **32**, 236–240.
1911. Trasformazione di una relazione funzionale dovuta al Dini. *R. C. Accad. Lincei*, (5) **20**, (i) 286–296, 381–391.
1911. Sulla espressione del resto in una operazione funzionale usata da Lord Rayleigh. *R. C. Accad. Lincei*, (5) **20**, (i) 605–614.

1911. Sur les équations générales du mouvement d'un corpuscule dans un champ magnétique et un champ électrique superposés. *Ark. Math. Naturvid. B*, **31,** no. 12, 1–7.
1911. Sur les équations lineaires à coefficients périodiques et sur le moyen mouvement du noeud lunaire. *Ann. École Norm. Sup.* (3) **28,** 325–376.
1911. Sullo spostamento dell' equilibrio. *Atti Ist. Veneto*, **71,** 241–249.
1912. Sulle onde di canale. *R. C. Accad. Lincei*, (5) **21,** (i) 3–14.
1912. Estensione ed evoluzione della fisica matematica nell' ultimo cinquantennio (con speciale riguardo al contributo italiano). *Atti. Soc. Ital. Progr. Sci.* **5,** 237–254 (Roma 1912). [Also in French, *Scientia*, **11,** 275–292.]
1912. Sulla continuazione analitica. *Atti Accad. Padova*, **28,** 61–63.
1912. Sulla gravitazione di un tubo sottile con applicazione all' anello di Saturno. *R. C. Circ. Mat. Palermo*, **33,** 354–374.
1912. Sir J. Larmor's mechanical model of the pressure of radiation. *Proc.* 5 *Congr. Int. Mat.* (Cambridge 1912), **1,** 217–220.
1912. Sur les systèmes linéaires à coefficients périodiques admettant une intégrale quadratique. *Ann. Sci. Acad. Polyt. Porto*, **7,** 193–206.
1913. Nuovo sistema canonico di elementi ellittici. *Ann. Mat. pura appl.* (3) **20,** 153–169.
1913. Sulla trasformazione delle equazioni lineari a derivate parziali del secondo ordine. *Atti Ist. Veneto*, **72,** 1331–1357.
1913. Sulle funzioni che ammetono una formula d'addizione del tipo $\sum_{1}^{n} X_i(x)\, Y_i(y)$. *R. C. Accad. Lincei*, (5) **22,** (ii) 181–183.
1913. Théorème de Torricelli et début de l'écoulement. *C. R. Acad. Sci. Paris*, **157,** 481–484.
1914. Sforzo di regime e sforzo d'avviamento per veicoli trainati. *Atti Ist. Veneto*, **73,** 931–946.
1914. Deduzione rigorosa di una relazione fondamentale nella teoria del calore raggiante. *R. C. Accad. Lincei*, (5) **23,** (i) 12–21.
1914. Sul regime variabile del calore raggiante. *R. C. Accad. Lincei*, (ii) 372–379, 453–464.
1915. Sulla riduzione del problema dei tre corpi. *Atti Ist. Veneto*, **74,** 907–939.
1915. Una proprietà di simmetria delle traiettorie dinamiche spiccate da due punti. *R. C. Accad. Lincei*, (5) **24,** (i) 666–674.
1915. Sulla regolarizzazione del problema piano dei tre corpi. *R. C. Accad. Lincei*, (5) **24,** (ii) 61–75.
1915. Forma mista di equazioni del moto che conviene ad una particolare categoria di sistemi meccanici. *R. C. Accad. Lincei*, (5) **24,** (ii) 236–248.
1915. Sul problema piano dei tre corpi. *R. C. Accad. Lincei*, (5) **24,** (ii) 421–433, 485–501, 553–569.
1916. Sulla introduzione di rincoli olonomi nelle equazioni dinamiche di Hamilton. *Atti Ist. Veneto*, **75,** 387–395.
1916. Sopra due trasformazioni canoniche desunte dal moto parabolico. *R. C. Accad. Lincei*, (5) **25,** (i) 445–458.
1916. Sur la régularisation du problème des trois corps. *C. R. Acad. Sci. Paris*, **162,** 625–628.
1917. Sulle linee d'azione degli ingranaggi. *Atti Accad. Padova*, **33,** 133–138.
1917. Sulle espressione analitica spettante al tensore gravitazionale nella teoria di Einstein. *R. C. Accad. Lincei*, (5) **26,** (i) 381–391.
1917. Nozione di parallelismo in una varietà qualunque e conseguente specifica zione della curvatura riemanniana. *R. C. Circ. Mat. Palermo*, **42,** 173–215.
1917. Statica einsteiniana. *R. C. Accad. Lincei*, (5) **26,** (i) 458–470.
1917. Realtà fisica di alcuni spazi normali del Bianchi. *R. C. Accad. Lincei*, (5) **26,** (i) 519–531.

1917/1919. $ds^2$ einsteiniani in campi newtoniani [note 1–9]. *R. C. Accad. Lincei*, (5) **26,** (ii) 307–317; **27,** (i) 3–12; **27,** (ii) 183–191, 220–229, 240–248, 283–292, 344–351; **28,** (i) 3–13, 101–109.

1918. La teoria di Einstein e il principio di Fermat. *Nuovo Cim.* (6) **16,** 105–114.

1919. Come potrebbe un conservatore giungere alla soglia della nuova meccanica. *R. C. Semin. Mat. Univ. Roma,* **5,** 10–28. [Spanish translation, *Rev. Mat. Hisp.-Amer.* **2,** 107–115, 123–132, 169–176; French translation, *Enseign. Math.* **21,** 5–28 (1920).]

1920. Sur la régularisation du problème des trois corps. *Acta Math. Stockh.* **42,** 99–144.

1920. Armonica viciniore ad una funzione assegnata. *R. C. Accad. Lincei,* (5) **29,** 197–206.

1920. L'ottica geometrica e la relatività generale di Einstein. *Riv. Ottica,* **1,** 187–200.

1922. Risoluzione dell' equazione funzionale che charatterizza le onde periodiche in un canale molto profondo. *Math. Ann. B,* **85,** 256–279.

1923. Sulla determinazione sperimentale dei coefficienti di un $ds^2$ einsteiniano. *Elettrotecnica,* **10,** 73–74.

1923. (With U. Amaldi). Lezioni di meccanica razionale. **1,** pp. xiii+741.

1923. Sulla velocità di trasporto nel moto ondoso permanente. *Scr. Univ. Bibl. Hierosolym.* **1,** no. 9, 1–14.

1923. Differenciales segundas que se compartan de modo invariantivo. *Rev. Mat. Hisp.-Amer.* **5,** 165–176.

1923. Sulla stabilità delle lavagne a cavalletto. *Period. Mat.* (4) **4,** 59–73.

1924. Questioni di meccanica classica e relativistica. Pp. 185. Zanichelli, Bologna. [Catalan edition, Inst. Estud Catalanis, pp. iv+151 (1922); German translation, pp. vi+110. Springer, Berlin.]

1924. Vorträge aus dem Gebiete der Hydro- und Aerodynamik (ed. Th. Kármán & T. Levi-Civita). Springer, Berlin.

1924. Determinazione rigorosa delle onde irrotazionali periodiche in acqua profonda. *R. C. Accad. Lincei,* (5) **33,** (ii) 141–150.

1925. Détermination rigoreuse des ondes permanentes d'ampleur finie. *Math. Ann. B,* **93,** 264–314. [Also, summarized, in *Proc. Int. Congr. Appl. Mech.* 1925, 129–145.]

1925. Lezioni di calcolo differenziale assoluto (edited by E. Persico). Pp. 315. Stock, Rome. [English edition (with two additional chapters on relativity), pp. xvi+350, Blackie, Glasgow; German edition, pp. xi+310, Springer, Berlin.]

1925. (With U. Amaldi). Condizioni atte ad assicurare l'indipendenza degli argomenti nella espressione hamiltoniana dell' azione variata. *R. C. Accad. Lincei,* (6) **1,** 263–272.

1925. Moti gravitazionali in una dimensione. *R. C. Accad. Lincei,* (6) **2,** 365–371.

1926. Commemorazione del Socio Nazionali Gregario Ricci-Curbastro. *Mem. Accad. Lincei,* **1,** fasc. 8, 555–567.

1926. Sullo scostamento geodetico. *Boll. Unione Mat. Ital.* **5,** 60–64.

1926. (With O. M. Corbino.) Il principio di Doppler e l'ipotesi balistica della luce. *R. C. Accad. Lincei,* (6) **3,** 644–, 705–.

1926. Sui moti einsteiniani in seconda approssimazione. *R. C. Accad. Lincei,* (6) **4,** 3–5.

1926/1927. (With U. Amaldi.) Lezioni di meccanica razionale. Vol. **2,** parte **1,** pp. ix+527; parte **2,** pp. ix+685. Zanichelli, Bologna.

1926. Sur l'écart géodesique. *Math. Ann. B,* **97,** 291–320.

1927. Sur les chocs dans le problème des trois corps. *C. R. 2 Congr. Int. Mec. Appl.* (Zürich), 96–106.

1928. With U. Amaldi.) Compendio di meccanica razionale. I, pp. xi+375; II, pp. viii+286. Zanichelli, Bologna.

1928. Fondamenti di meccanica relativistica [translated by E. Persico]. Pp. vii+185.

1928. Drei Vorlesungen ueber adiabatische Invarianten. *Abh. Math. Sem. Hamburg Univ. B*, **6,** 323–366.

1928. Sul massimo cimento dinamico nei sistemi elastici. *R. C. Sem. Mat. Fis. Milano*, **2,** 78–109. [Polish translation, *Przegl. Tech.* (1930).]

1928. Sul moto di un corpo di massa variabile. *R. C. Accad. Lincei*, (6) **8,** 329–333, 621–622.

1928. Sugli invarianti adiabatici. *Atti Congr. Int. Fis.* (Como 1927), 475–513.

1928. Alcune applicazioni astronomiche degli invarianti adiabatici. *Boll. Unione Mat. Ital.* **7,** 251–253.

1929. Dynamische Beanspruchung elastischer Systeme. *Mh. Math. Phys. B*, **36,** 165–192.

1929. A proposed modification of Einstein's field-theory. *Nature, Lond.* **123,** 678–679.

1929. Vereinfachte Herstellung der Einsteinschen einheitlichen Feldgleichungen. *S. B. preuss. Akad. Wiss.* **9,** 137–153. [English translation, Blackie, Glasgow.]

1929. Osservazione (added to a note by P. Nalli, Sopra una spostamento generalizzato negli spazi di Riemann). *R. C. Accad. Lincei*, (6) **9,** 960–962.

1929/1930. (With G. Fubini.) Sulle curve analoghe al circolo osculatore quando si passa da tre a quattro punti infinitamente vicini. *Ann. Mat. pura appl.* (4) **7,** 193–211.

1930. With U. Amaldi.) Lezioni di meccanica razionale. Ed. 2, Vol. **1,** pp. xiv+897. Zanichelli, Bologna.

1930. Caratteristiche e bicaratteristiche della equazioni gravitazionali di Einstein. *R. C. Accad. Lincei*, (6) **11,** 3–11, 113–121.

1930. Corpuscoli cosmici e distribuzione maxwelliana. *Atti Accad. Nuovi Lincei*, **83,** 176–189.

1930. Ancora sul moto di un punto di massa variabile. *R. C. Accad. Lincei*, (6) **11,** 626–632.

1930. Allgemeine Folgerungen aus der Prandtlschen Grenzschichttheorie. *In* Vorträge aus dem Gebiete der Aerodynamik, 30–50. Springer, Berlin.

1930. Sui potenziali newtoniani dovuti a distribuzioni che si estendono illimitatamente. *R. C. Sem. Univ. Padova*, **1,** 133–156. [Spanish translation, *Rev. Mat. Hisp.-Amer.* (2) **6,** 65–85 (1931).]

1930. Sezioni piane di un corpo e direttrici ortopariche. *R. C. Accad. Lincei*, (6) **12,** 535–541.

1931. Commemorazione del Socio Straniero Paul Appell. *R. C. Accad. Lincei*, (6) **13,** 241–242.

1931. Applicazioni astronomiche degli invarianti adiabatici. *Atti Congr. Int. Mat.* (1928), **5,** 17–28.

1931. A proposito delle Note dei sig.-ri Hatzidakis e Sakellariou sui moti centrali. *R. C. Accad. Lincei*, (6) **13,** 715–718.

1931. Rifrazione e riflessione nella relatività generale. *Atti Accad. Nuovi Lincei*, **84,** 332–352.

1931. Sui getti liquidi. *R. C. Sem. Mat. Fis. Milano*, **5,** 154–173. [Polish translation, *Czas. Tech. Lwów* (1933).]

1931. Ueber Zermelo's Luftfahrtproblem. *Z. angew, Math. Mech. B*, **11,** 314–322.

1931. Sur les surfaces admettant un réseau triangulaire de lignes parallèles. *C. R. Acad. Sci. Paris*, **193,** 1171–1173.

1931. Caratteristiche dei sistemi differenziali e propagazione ondosa (edited by G. Lamparello). Pp. vii+105. Zanichelli, Bologna. [French edition, pp. x+114, Alcan, Paris.]

1932. Attrazione newtoniana dei tubi sottili e vortici filifomi. *Ann. Scu. Norm. Sup. Pisa*, (2) **1,** 1–33, 229–250.

1932. Sur les jets liquides. *J. Math. pures appl.* **11,** 37–56.

1932. Teoremi di unicità e di esistenza per la piccole oscillazioni di un filetto vorticoso prossimo alla forma circolare. *R. C. Accad. Lincei,* (6) **15,** 409–416.
1932. Sopra un probleme di navigazione aerea (rotta entro un campo di vento). *Aerotecnica,* **12,** 725–727.
1933. Diracsche und Schrödingsche Gleichungen. *S. B. preuss. Acad. Wiss.* **5,** 240–250.
1933. Ergänzende Bemerkungen zum Weierstraszschen Vorbereitungssatz und bedingt-periodische Bewegungen. *Z. angew. Math. Mech. B,* **13,** 112–114.
1933. Some mathematical aspects of the new mechanics. *Bull. Amer. Math. Soc.* **39,** 535–563.
1934. Terne di congruenze sopra una superficie ed estensione della trigonometria. *Compositio Math.* **1,** 115–162.
1934. A general survey of the theory of adiabatic invariants. *J. Math. Phys.* **13,** 18–40.
1934. The secular effects of tides on the motion of planetary systems. *Amer. Math. Mon.* **41,** 279–296.
1934. Sulle soluzioni stazionarie dei sistemi pfaffiani. I, Generalità e richiani. II, Il caso più significativo. *R. C. Accad. Lincei,* (6) **19,** 261–267, 369–375.
1934. Alcuni aspetti matematici della nuova meccanica. *Nuovo Cim.* n.s. **11,** 173–200.
1934. Curve chiuse a parallelismo monodromo sopra la sfera. *Atti Accad. Nuovi Lincei,* **87,** 407–414.
1934. Perfezionamente della regola di equivalenza fra moti einsteiniani e moti newtoniani. *R. C. Accad. Lincei,* (6) **20,** 398–403.
1935. (With U. Amaldi.) Teoria matematica dei fenomini ondosa. *Encic. Ital.* **25,** art. Onde.
1935. Movimenti di un sistema continuo che rispettano l'invariabilità sostenziale del baricentro. *Atti Accad. Nuovi Lincei,* **88,** 151–155.
1935. (With U. Amaldi.) Nozioni di balistica esterna. Pp. viii+56. Zanichelli, Bologna.
1935. Points matériels et corps célestes en relativité générale—Etude préliminaire. In the volume presented to M. Marcel Brillouin.
1936. Le problème des deux corps en relativité générale. *Enseign. Math.* **34,** 149–175.
1936. Sulla nozione di intervallo fra due avvenimenti: primo approcio alla relatività. *Nuovo Cim.* n.s. **13,** 45–65.
1936. Movimenti per sola gravitazione di un sistema continuo. *Scritti matematici offerti a Luigi Berzolari,* 240–247. Pavia.
1936. Sur une transformation de la fonction lagrangienne qui n'altère pas les trajectoires. *Recueil Math. Moscou,* n.s. **1,** 123–125.
1937. The relativistic problem of several bodies. *Amer. J. Math.* **59,** 9–22.
1937. Astronomical consequences of the relativistic two body problem. *Amer. J. Math.* **59.**
1937. Forme canoniche dei $ds^2$ binari con data curvatura totale. *R. C. Accad. Lincei,* (6) **25,** 197–205.

# Selected References


[1] Accademia Nazionale dei Lincei. Tullio Levi-Civita. *Convegno internazionale celebrativo del centenario della nascita: Roma, 17–19 dicembre 1973. Rome*, 1975.

[2] U. Amaldi, *Commemorazione del socio Tullio Levi-Civita* (Italian), Atti Accad. Naz. Lincei. Rend. Cl. Sci. Fis. Mat. Nat. (8) **1** (1946), 1130–1155. MR0020061

[3] Max Born and Max von Laue, "Max Abraham." *Physikalische Zeitschrift,* 24 (1923): 49–53.

[4] R. J. B. Bosworth, *Mussolini's Italy: Life under the Fascist Dictatorship, 1915–1945.* New York, 2007.

[5] Umberto Bottazzini, *Ricci and Levi-Civita: from differential invariants to general relativity,* The symbolic universe (Milton Keynes, 1996), Oxford Univ. Press, New York, 1999, pp. 241–259. MR1750739

[6] Carl B. Boyer, *A history of mathematics,* John Wiley & Sons, Inc., New York-London-Sydney, 1968. MR0234791

[7] Carlo Cattani and Michelangelo De Maria, *Max Abraham and the reception of relativity in Italy: his 1912 and 1914 controversies with Einstein,* Einstein and the history of general relativity (North Andover, MA, 1986), Einstein Stud., vol. 1, Birkhäuser Boston, Boston, MA, 1989, pp. 160–174. MR1200721

[8] ———, "The 1915 Epistolary Controversy between Einstein and Tullio Levi-Civita." In Don Howard and John Stachel, eds., *Einstein and the History of General Relativity,* 175–200. Boston, 1989.

[9] Emanuele Cenghiaro and Pier Giovanni Zanetti. *Padova e la Grande Guerra.* Padua, 2008.

[10] Claudio Citrini. "Matematica e vita civile nel Politecnico di cento anni fa: la vicende di Max Abraham." *Annali di Storia delle Università* Italiane 12 (2008): 101–117.

[11] Umberto Cisotti and Carlo Somigliana. "Tullio Levi-Civita." *R. Istituto Lombardo di Scienze, lettere Arti. Rendiconti,* 75 (1942): 110–112.

[12] Salvatore Coen (ed.), *Mathematicians in Bologna 1861–1960,* Birkhäuser/Springer Basel AG, Basel, 2012. MR2934582

[13] *The Collected Papers of Albert Einstein,* 14 vols. Princeton, 1987–2015.

[14] Freeman J. Dyson, *Unfashionable pursuits,* Math. Intelligencer **5** (1983), no. 3, 47–54. MR737690

[15] Albert Einstein, *Autobiographical notes,* A bilingual German/English edition, Open Court Publishing Co., La Salle, Ill., 1979. Translated from the German and edited by Paul Arthur Schilpp. MR529428

[16] ———, *Relativity: The Special and General Theory.* transl. R. W. Lawson, New York, 1921.

[17] Luther Pfahler Eisenhart, *Riemannian Geometry,* Princeton University Press, Princeton, N. J., 1949. 2d printing. MR0035081

[18] ———, "Ricci's Principal Directions for a Riemann Space and the Einstein Theory." *Proceedings of the National Academy of Sciences* 8 (1922): 24–26.

[19] Jean Eisenstaedt, *The curious history of relativity*, Princeton University Press, Princeton, NJ, 2006. How Einstein's theory of gravity was lost and found again; Translated from the 2003 French original by Arturo Sangalli; With a foreword by Thibault Damour. MR2264671

[20] Philipp Frank, *Einstein: His Life and Times,* transl. G. Rosen, New York, 1947.

[21] Peter Louis Galison, "Minkowski's Space-Time: From Visual Thinking to the Absolute World." *Historical Studies in the Physical Sciences* (1979): 85–121.

[22] Livia Giacardi and Rossana Tazzioli, eds., *Le lettere di Eugenio Beltrami a Betti, Tardy and Gherardi.* Milan, 2012.

[23] Stanley Goldberg, *Understanding relativity*, Birkhäuser Boston, Inc., Boston, MA, 1984. Origin and impact of a scientific revolution. MR736684

[24] Judith R. Goodstein, "Levi-Civita, Einstein, and Relativity in Italy." *Atti Rendiconti Accademia dei Lincei* 8 (1975): 43–51.

[25] ———, "The Italian Mathematicians of Relativity." *Centaurus* 26 (1983): 241–261.

[26] ———, *The Volterra Chronicles: The Life and Times of an Extraordinary Mathematician, 1860–1940.* AMS, Providence, 2007.

[27] *Gregorio Ricci Curbastro: La vita di un liceo e l'opera di un matematico,* C. Bartolotti and A. Pirazzini, eds., Lugo, 2002.

[28] Hanoch Gutfreund, and Jürgen Renn. *The Road to Relativity: The History and Meaning of Einstein's "The Foundation of General Relativity."* Princeton, 2015.

[29] Walter Isaacson. *Einstein: His Life and Universe.* New York, 2007.

[30] Michel Janssen and Jürgen Renn, eds., "Arch and Scaffold: How Einstein Found his Field Equations." *Physics Today* 68 (2015): 30–36.

[31] Louis Kollros. "Albert Einstein en Suisse. Souvenirs." *Helvetica Physica Acta Supplement* 4 (1956): 271–281.

[32] Tullio Levi-Civita, *Opere matematiche. Memorie e note. Vol. I. 1893–1900* (Italian), Nicola Zanichelli Editore, Bologna, 1954. Pubblicate a cura dell'Accademia Nazionale dei Lincei. MR0062680

[33] ———, *Lezioni di calcolo differenziale assoluto,* ed. Enrico Persico, Rome, 1925; *The Absolute Differential Calculus (calculus of tensors),* ed. E. Persico, Transl. M. Long, London, 1927.

[34] ———, "Commemorazione del Socio Nazionale Gregorio Ricci-Curbastro." *Mem. Accad. Lincei* 1 (1926): 555–567.

[35] Sandra Linguerri and Raffaella Simili, eds. *Einstein parla italiano: Itinerari e polemiche.* Bologna, 2008.

[36] MacTutor History of Mathematics archive. Available at http://www-history.mcs.st-andrews.ac.uk/index. html.

[37] Alberto Malfitano, Gian D. Veggi, and Antonio Curzi. *1859: la Romagna e Lugo dalle Legazioni al Regno d'Italia.* Lugo di Romagna, 2010.

[38] Pietro Nastasi and Rosanna Tazzioli. *Aspetti scientifici e umani nella corrispondenza di Tullio Levi-Civita (1873–1941).* Palermo, 2000.

[39] ———, "Toward a Scientific and Personal Biography of Tullio Levi-Civita (1873–1941)." *Historia Mathematica* 32 (2005): 203–236.

[40] John Norton, *How Einstein found his field equations, 1912–1915*, Einstein and the history of general relativity (North Andover, MA, 1986), Einstein Stud., vol. 1, Birkhäuser Boston, Boston, MA, 1989, pp. 101–159. MR1200720

[41] ———, "A Peek into Einstein's Zurich Notebook." Available at `http://www.pitt.edu/~jdnorton/Goodies`.

[42] Abraham Pais, *"Subtle is the Lord..."*, The Clarendon Press, Oxford University Press, New York, 1982. The science and life of Albert Einstein. MR690419

[43] Karin Reich, *The American contribution to the theory of differential invariants, 1900–1916*, The attraction of gravitation: new studies in the history of general relativity

(Johnstown, PA, 1991), Einstein Stud., vol. 5, Birkhäuser Boston, Boston, MA, 1993, pp. 225–247. MR1735383

[44] Gregorio Ricci-Curbastro, *Opere. Note e Memorie,* 2 vols. Rome, 1956.

[45] ———, "Direzioni e invarianti principali in una varietà qualunque." *Atti del Reale Istituto Veneto di Scienze, Lettere ed Arti* 63 (1904): 1233–1239.

[46] M. M. G. Ricci and T. Levi-Civita, *Méthodes de calcul différentiel absolu et leurs applications* (French), Math. Ann. **54** (1900), no. 1-2, 125–201. MR1511109

[47] Jürgen Renn, "The Summit Almost Scaled: Max Abraham as a Pioneer of a Relativistic Theory of Gravitation." In J. Renn, ed., *The Genesis of General Relativity,* vol. 3, 1229–1255. Dordrecht, 2007.

[48] David E. Rowe, *Einstein's gravitational field equations and the Bianchi identities*, Math. Intelligencer **24** (2002), no. 4, 57–66. MR1930868

[49] H. S. Ruse, *Obituary: Tullio Levi-Civita*, Edinburgh Math. Notes **33** (1943), 19–24. MR0008797

[50] Tilman Sauer, "Albert Einstein, Review Paper on General Relativity Theory (1916)." In I. Grattan-Guinness, ed., *Landmark Writings in Western Mathematics, 1640–1940,* 802–822. Amsterdam, 2005.

[51] ———, "Marcel Grossmann and his contribution to the general theory of relativity." [arXiv:1312.4068v2 [physics.hist-ph]

[52] Carl Seelig, *Albert Einstein: A Documentary Biography,* transl. M. Savill, London, 1956.

[53] Francesco Severi, "Discorso." In *Celebrazione in Lugo del centenario della nascita di Gregorio Ricci Curbastro,* 55– 70. Lugo, 1954.

[54] Carlo Somigliana, *Obituary: Tullio Levi-Civita e Vito Volterra* (Italian), Rend. Sem. Mat. Fis. Milano **17** (1946), 1–15. MR0024376

[55] Dirk J. Struik, "Outline of a History of Differential Geometry." *Isis* 20 (1933): 161–191.

[56] ———, *A Concise History of Mathematics.* New York, 1967.

[57] ———, "J. A. Schouten and the Tensor Calculus." *Nieuw Archief Voor Wiskunde* 26 (1978): 96–107.

[58] Dirk J. Struik, *Schouten, Levi-Civita, and the emergence of tensor calculus*, The history of modern mathematics, Vol. II (Poughkeepsie, NY, 1989), Academic Press, Boston, MA, 1989, pp. 99–105. MR1037810

[59] ———, "From Riemann to Ricci: The Origins of the Tensor Calculus." In H. M. Srivastava and Th. M. Rassias, eds., *Analysis, Geometry and Groups: A Riemann Legacy Volume,* 657–674. Palm Harbor, 1993.

[60] Tina Tomasi and Nella Sistoli Paoli, *La Scuola normale di Pisa dal 1813 al 1945: cronache di un'istituzione.* Pisa, 1990.

[61] Fabio Toscano, *Il genio e il gentiluomo: Einstein e il matematico italiano che salvò la teoria della relatività generale.* Milan, 2004.

[62] ———, "Luigi Bianchi, Gregorio Ricci Curbastro e la scoperta delle identità di Bianchi." In *Atti del XX Congresso Nazionale di Storia della Fisica e dell'astronomia,* ed. E. Schettino, 353–370. Naples, 2001.

[63] Angelo Ventura, *Padova.* Bari, 1989.

# Notes

## Abbreviations

| | |
|---|---|
| ACS | Archivio Centrale dello Stato (Rome) |
| ASR | Archivio di Stato, Roma |
| ASUR | Archivio Storico dell'Università di Roma |
| CPAE | The Collected Papers of Albert Einstein |
| MI | Ministero dell'Interno |
| MPI | Ministero della Pubblica Istruzione–Direzione Generale Istruzione Superiore |
| SUB Göttingen | Niedersächsische Staats- und Universitatsbibliothek Göttingen |
| UIS | Università e Istruzione Superiore |
| DGIU | Direzione Generale Istruzione Universitaria |

## Notes for Chapter 1

1. Gregorio Ricci Curbastro [surname hereafter given as Ricci], *Suor Vincenza Ricci Curbastro, Figlia della Carità* (Padua, 1896), 6.

2. Quoted in Fabio Toscano, *Il genio e il gentiluomo: Einstein e il matematico italiano che salvò la teoria della relativita generale* (Milan, 2004), 129.

3. G. Ricci, *Suor Vincenza,* 6.

4. Francesco Taglioni, "Attestato di studio," October 15, 1869, ASR, Fondo Università, busta 590, fasc. 7700.

5. In 1935, the university moved to its present site near the train station.

6. Denis Mack Smith, *Modern Italy: A Political History,* rev. ed. (Ann Arbor, 1969), 86.

7. Charles Dickens, *Pictures from Italy,* edited with notes and an introduction by Kate Flint (London, 1998), 150.

8. Gaspare Finali to Cesare Correnti, October 24, 1871, ACS, MPI, Personale, busta 1788, fasc. Gregorio Ricci.

9. Dora Dumont, "The Nation As Seen from Below: Rome in 1870," *European Rev. of History* 15 (2008), 489.

10. G. Ricci to Angelo Manzoni, November 24, 1872, quoted in F.Toscano, *Il genio,* 134.

## Notes for Chapter 2

1. Quoted in Giovanni Sansone, *Algebristi, analisti, geometri differenzialisti, meccanici e fisici-matematici ex-normalisti del periodo 1860-1929* (Pisa, 1977), 12.
2. Vito Volterra, Senato del Regno, Atti parlamentari. Discussioni. November 22, 1918.
3. G. Ricci to Alfredo Capelli, March 17, 1895, Fondo Ricci Curbastro, Biblioteca Trisi, Lugo.
4. Ibid.
5. G. Ricci, "Cenno sulla vita scientifica del Dott. Gregorio," February 18, 1880, ASR, Fondo Università, busta 1089 (miscellanea), fasc. documenti e pubblicazione relativi all'assistente universitaria Gregorio Ricci.
6. An undated document listed as #52 in the inventory of the Ricci papers preserved in the Liceo Scientifico di Lugo.
7. Enrico Betti, "Attestato," March 20, 1877, ASR, Fondo Università, busta 590, fasc. 7700.
8. "Sopra un sistema di due equazioni differenziali lineari di cui l'una è quella dei fattori integranti dell'altra" (1877).
9. "Sulla teoria elettrodinamica di Maxwell" (1877); "Sopra la deduzione di una nuova legge fondamentale di elettrodinamica" (1877).
10. An undated document, item #52, Ricci papers, Liceo Scientifico di Lugo.

## Notes for Chapter 3

1. E. Betti to unnamed official, July 18, 1876, ACS, MPI, Personale, busta 1788.
2. Ibid.
3. G. Ricci, *Suor Vincenza,* 8.
4. Domenico Ricci to G. Ricci, undated, Fondo Ricci Curbastro, Biblioteca Fabrizio Trisi, Lugo.
5. J. Plücker, *Neue Geometrie des Raumes, gegründet auf die Betrachtung der geraden Linie als Raumelement, mit einem Vorwort von A. Clebsch* (Leipzig, 1868-69).
6. David E. Rowe, "Klein, Hilbert, and the Göttingen Mathematical Tradition," *Osiris* 5 (1989), 192.
7. Felix Klein, "Vergleichende Betrachtungen über neuere geometrische Forschungen," vol. 1 of *Gesammelte mathematische Abhandlungen* (Berlin, 1921), 460-497.
8. F. Klein, *Development of Mathematics in the 19th Century,* trans. M. Ackerman (Brookline, 1979), 155.

9. The official name of the institution was the *Königlich Bayerische Technische Hochschule München*; in 1901, it gained the right to award doctorates and in 1970 changed its name to the Technische Universität München.

10. Quoted in Livia Giaccardi, "Models in Mathematical Teaching in Italy (1850-1950)," http://math-art.eu/Cagliari2013-Lectures.php. See also, D. Rowe, "Mathematical models as artefacts for research: Felix Klein and the case of Kummer surfaces," *Mathematische Semesterberichte,* 60 (2013), 1–24.

11. The 1878-1879 academic year in Munich consisted of only two semesters: a winter semester that ended in March, and a summer semester that began after Easter and ended in July.

12. G. Ricci to E. Betti, December 29, 1878, Betti Archive, Archivio della Scuola Normale Superiore di Pisa.

13. Ibid.

14. G. Ricci, "Relazione di Gregorio Ricci Curbastro sul perfezionamento svolto a Monaco di Baviera presso il prof. Klein," Mar. 14, 1879, ACS, MPI, Personale, busta 1788.

15. F. Klein, "Lettera del prof. Klein," March 13, 1879, ACS, MPI, Personale, b. 1788.

16. F. Klein, *Development of Mathematics,* 348. His papers on the subject were published in 1878 and 1879 in *Math. Ann.*

17. G. Ricci, "Relazione semestrale di G. Ricci," ACS, MPI, Personale, busta 1788.

18. F. Klein, "Lettera del prof. Klein," July 29, 1879, ibid.

19. F. Klein, *Development of Mathematics,* 278.

20. G. C. Young, "Professor Klein," *The Times,* 1925.

21. G. Ricci to F. Klein, November 3, 1892, SUB Göttingen, Cod. Ms. F. Klein 11:500.

## Notes for Chapter 4

1. Ulisse Dini to F. Mazzuoli, November 12, 1879, ACS, MPI, Personale, busta 1788.

2. U. Dini to unknown correspondent, Nov. 13, 1879, ibid.

3. G. Ricci to E. Betti, December 3, 1880, Betti Archive, Scuola Normale Superiore di Pisa.

4. E. Betti to G. Ricci, undated draft, ibid.

5. V. Volterra, "Betti, Brioschi, Casorati: Tre analisti e tre modi di considerare le questioni d'analisi," *Saggi scientifici* (Bologna, 1920; reprint, with an introduction by Raffaella Simili, Bologna, 1990), 49.

6. G. Ricci, "Origini e sviluppo dei moderni concetti fondamentali sulla geometria," *Opere,* vol. 2 (Rome, 1956-57), 292.

7. Quoted in Tullio Levi-Civita, "Gregorio Ricci Curbastro," *Opere,* vol. 1, 1.

8. F. Klein, *Development of Mathematics,* 203.

9. T. Levi-Civita, "Commemorazione del socio nazionale prof. Gregorio Ricci Curbastro," Mem. Acc. Lincei 1(1926), 557.

10. D. Ricci to G. Ricci, January 20, 1881, Fondo Ricci Curbastro, Biblioteca, Lugo.

11. T. Levi-Civita, "Commemorazione," 561.

12. Angelo Tonolo, "Commemorazione di Gregorio Ricci Curbastro nel primo centenario della nascita," *Rendiconti del Seminario della Università di Padova,* 23 (1954), 19.

13. Ibid.

14. Maria Pastori, "Necrologio di Angelo Tonolo," *Bollettino dell'Unione Matematica Italiana,* 17 (1962), 423.

15. Livia Giacardi, "Federigo Enriques (1871-1946) and the training of mathematics teachers in Italy," in Salvatore Coen, ed., *Mathematicians in Bologna 1861-1960* (Basel, 2012), 211.

16. Giuseppe Bettamini *et al* to "Onoverole Sig. Rettore," December 11, 1882, ACS, MPI, Personale, busta 1788.

17. Terenzio Mamiani to Guido Baccelli, January 27, 1883, ibid.

18. Francesco Brioschi to Michele Coppino, November 26, 1884, ibid.

19. Aldo Finzi to T. Levi-Civita, May 19, 1926, Levi-Civita Personal Papers, courtesy of Ceccherini-Silberstein family.

## Notes for Chapter 5

1. Antonio Ricci to G. Ricci, July 27, 1879, Fondo Ricci Curbastro, Biblioteca di Lugo.

2. Bianca Bianchi Azzarani to G. Ricci, January 5, 1884, Fondo Ricci Curbastro, Biblioteca di Lugo.

3. Ibid. January 20, 1884.

4. Ibid. January 18, 1884.

5. Ibid.

6. Ibid. February 5, 1884.

7. Quoted by Dirk Struik, in "From Riemann to Ricci: The Origins of the Tensor Calculus," H. M. Srivastava, T. M. Rassias, eds., *Analysis, Geometry, and Groups: A Riemann Legacy* (Palm Harbor, 1993), 657.

8. Bernhard Riemann, "Über die Hypothesen welche der Geometrie zu Grunde liegen" ("On the Hypotheses That Lie at the Foundations of Geometry").

9. B. Riemann, "Über eine Frage der Wärmeleitung" ("On a Question of Heat Conduction").

10. Olivier Darrigol, "The Mystery of Riemann's Curvature," *Historia Mathematica* 42 (2015), 48.

11. F. Klein, *Development of Mathematics,* 237.

12. G. Ricci, "Principii di una Teoria delle Forme Differenziali Quadratiche" ("Principles of a theory of differential quadratic forms"), *Opere,* vol. 1, 140.

13. B. Bianchi to G. Ricci, March 18, 1884, Fondo Ricci Curbastro, Biblioteca di Lugo.

## Notes for Chapter 6

1. *La Venezia,* 24 February 1887.

2. T. Levi-Civita, "Commemorazione," 560.

3. Francesco G. Tricomi, "Matematici italiani del primo secolo dello stato unitario," *Mem. Acc. Sci. Torino,* Classe di Scienze Fisiche, Matematiche e Naturali, 4th ser. 1 (1962), 41.

4. Giuseppe Veronese to M. Coppino, July 8, 1884, ACS, MPI, DGIS, UIS, busta 563, fasc. 510, sottofasc. 4.

5. Luigi Cremona, August 1886, ibid.

6. Quoted in "Seduta straordinaria, dei soli Professori Ordinari, 9 Novembre 1886," ibid. All details of this meeting, unless otherwise noted, are from this source.

7. See Giovanni Canestrini to Gianpaolo Vlacovich, October 26, 1886; G. Vlacovich to M. Coppino, October 30, 1886; and G. Veronese to Commendatore [unidentified], November 2, 1886, ibid.

8. U. Dini to M. Coppino, December 5, 1886, ACS, MPI, DGIS, UIS, busta 465, fasc. 23, sottofasc. 4.

9. G. Ricci to M. Coppino, December 3, 1886, ibid.

10. Giuseppe Lorenzoni, Giovanni Omboni, Ernesto Padova, Francesco D'Arcais, Augusto Righi, to M. Coppino, Feb. 28, 1887, ACS, MPI, UIS, busta 563, fasc. 510, sottofasc. 4.

11. G. Canestrini to G. Vlacovich, March 3, 1887, ibid.

12. Ibid.

13. G. Ricci to M. Coppino. Dec. 3, 1886, ACS, MPI, DGIS, UIS, busta 465, fasc. 23, sottofasc. 4.

14. G. Ricci to M. Coppino, June 19, 1887, ACS, MPI, DGIS, UIS, busta 563, fasc. 510, sottofasc. 4.

15. U. Dini to Giovanni Ferrando, June 18, 1887, ibid.

16. G. Ricci to V. Volterra, May 15, 1888, Vito Volterra Papers, Accademia Nazionale dei Lincei, Rome.

## Notes for Chapter 7

1. T. Levi-Civita, "Commemorazione," 561.
2. It came with a cash prize of 10,000 *lire* to be awarded annually for the best papers or discoveries in the physical, mathematical, and natural sciences, or the moral, historical, and philological sciences.
3. In private, Beltrami described the fourth entry, submitted by G. Ribaldi, an engineer, as trivial and full of errors, and barely mentioned it in his report.
4. Eugenio Beltrami to E. Betti, April 12, 1889, in L. Giacardi and R. Tazzioli, eds., *Le lettere di Eugenio Beltrami a Betti, Tardy e Gherardi* (Milan, 2012), 163.
5. E. Beltrami to E. Betti, May 9, 1889, *Le lettere,* 164.
6. E. Beltrami to E. Betti, June 11, 1889, ibid., 165.
7. "Relazione sul concorso al premio Reale per la Matematica per l'anno 1887," *Atti della R. Accademia dei Lincei* 5 (1889), 307.
8. E. Beltrami, "Saggio di interpretazione della geometria non-euclidean," *Giornale di matematiche* 6 (1868), 284-312.
9. D. Struik, "From Riemann to Ricci," 666.
10. G. Ricci, "Principii di una teoria delle forme differenziali quadratiche," *Opere,* vol. 1, 139–140.
11. E. Beltrami, "Relazione sul concorso al premio Reale per la Matematica per l'anno 1887," 304. For an illuminating study of how the Italian mathematical community viewed Ricci's work, see also Umberto Bottazzini, "Ricci and Levi-Civita: From Differential Invariants to General Relativity," in Jeremy Gray, *The Symbolic Universe: Geometry and Physics 1890-1930* (Oxford, 1999), 241–258.
12. E. Beltrami, 305.
13. G. Ricci, "Sui parametri e gli invarianti delle forme differenziali quadratiche," *Opere,* vol. 1, 177–188.
14. Quoted in U. Bottazzini, "Ricci and Levi-Civita," 244.
15. F. Toscano, *Il genio e il gentiluomo,* 164.
16. For a more detailed exposition, see Appendix A, page 133.
17. "Sulla derivazione covariante ad una forma quadratica differenziale," *Opere,* vol. 1, 199–203.
18. Quoted in U. Bottazzini, "Ricci and Levi-Civita," 245–246.
19. Quoted in T. Levi-Civita, "Commemorazione," 558.
20. *J. Reine Angew. Math.* 70 (1869), 46–70.
21. Quoted in O. Darrigol, "The Mystery of Riemann's Curvature," 70.

22. T. Levi-Civita, *Lezioni di calcolo differenziale assoluto* (Rome, 1925), vii.

23. D. Struik, "From Riemann to Ricci," 663.

24. Ibid., 665.

25. Luca Dell'Aglio, "On the Genesis of the Concept of Covariant Differentiation," *Revue d'histoire des mathématiques* 2 (1996), 255.

26. G. Ricci, "Delle derivazioni covarianti," 1888, 245. The independence of the form of equations with respect to changes in coordinate systems (which Ricci mentioned several times) was central to Einstein's adopting differential geometry for general relativity. In special relativity, Einstein managed to make the equations of motion (and of electromagnetism) invariant in form for the class of observers moving with constant velocity. (This solved the ether problem, because the speed of light emerged from the theory as invariant for all these observers.)

    In general relativity, Einstein sought to make equations invariant in form for any observer, including accelerated ones. This was later known as the principle of general covariance. (A more mature understanding of general covariance shows that it does not really cover a general class of observers, since these cannot really be identified with coordinate systems in general. But what general covariance does is to establish an even more important principle of general relativity—that there is no prior space-time geometry, but rather that it arises naturally from the solutions of the Einstein field equations.)

27. G. Ricci, "Delle derivazioni covarianti," *Opere,* vol. 1, 247. This is actually standard procedure in general relativity today. It appears everywhere in the Charles Misner, Kip Thorne, and John Wheeler textbook on gravitation, for instance: choose a convenient reference frame (such as a "local rest frame") to write equations conveniently using tensors, then elevate them automatically to general coordinates because their form does not change.

28. E. Beltrami, "Relazione," 306–307.

29. Ibid., 246.

30. Ibid.

31. Francesco Porro to T. Levi-Civita, May 5, 1926, Levi-Civita Personal Papers.

32. Valentino Cerruti, "Relazione sul concorso al Premio Reale per la Matematica per l'anno 1889," *Rendiconti delle sedute solenni della R. Accademia dei Lincei del 5 giugno 1892,* vol. 8, 40.

33. G. Ricci, "Di alcune applicazioni del calcolo differenziale assoluto alla teoria delle forme differenziali quadratiche binarie e dei sistemi a due variabili," 1893, *Opere,* vol. 1, 311.

34. T. Levi-Civita, "Commemorazione," 557–558.

## Notes for Chapter 8

1. Angelo Ventura, *Padova* (Bari, 1989), 120–121.

2. Mariarosa Davi, "I duecento anni del liceo Tito Livio," Padova e il suo territorio, 27, no. 160 (2012), 25-29. For a full account of his career, see Davi, "Giacomo Levi-Civita amministratore cittadino," M. Davi and Giulia Simone, eds., *Giacomo Levi-Civita e l'ebraismo veneto tra Otto e Novecento* (Padua, 2015), 15-34.

3. Information about Giacomo Levi-Civita's early life also draws on documents in the Archivio Storico del Senato della Repubblica, Rome, and from an unpublished interview by the author with Susanna Silberstein Ceccherini, Caprino Veronese, July 18, 2009. [Hereafter referred to as S.S.C. Interview]

4. Annie Sacerdoti and Luca Fiorentino, *Guida all'Italia ebraica* (Genoa, 1986), 119. See also, Ariel Viterbo, "Dal primo dopoguerra alla Shoà," in Claudia De Benedetti, ed., Il cammino della Speranza. Gli ebrei e Padova (Padua, 2000), vol. II, 133; Francesco Selmin, *Nessun "giusto per Eva."* La Shoah a Padova e nel Padovano (Sommacampagna, 2011).

5. Università La Sapienza, Archivio Storico, Tullio Levi-Civita, AS 487, no. 03, "R. Università degli Studi di Roma. Stato matricolare," undated.

6. For the complete list of courses, see files in the Liceo Tito Livio archives, "Registro della quinta ginnasio, 1886–1887" and "Elenco degli alunni premiati, 1886–1887," busta Atti 1886–1887.

7. Felix J. D. Carli to T. Levi-Civita, October 12, 1930, Tullio Levi-Civita Papers, Accademia dei Lincei, Rome.

8. Ugo Amaldi, "Commemorazione del Socio Tullio Levi-Civita," *Rendiconti dell'Accademia Nazionale dei Lincei,* ser. 7, vol. 1 (1946), 1135.

9. S.S.C. Interview.

10. E-mail from M. Davi to the author, August 2, 2014.

11. U. Amaldi, "Commemorazione," 1135.

12. William Valance Douglas Hodge, "Tullio Levi-Civita, 1873–1941," *Obituary Notices of Fellows of the Royal Society of London,* 4 (1942), 152.

13. W. V. D. Hodge, "Tullio Levi-Civita," 157.

14. Ibid., 152.

15. Adolfo Bartoli to Paolo Boselli, April 27, 1890, ACS, MPI, DGIS, UIS, busta 563, fasc. 797, sottofasc. 8.

16. Memorandum, A. Bartoli to P. Boselli, December 11, 1889, ibid.

17. G. Ricci to V. Volterra, Jan. 9, 1890, and V. Volterra to G. Ricci, Jan. 11, 1890, Vito Volterra Collection, Accademia Nazionale dei Lincei, Rome.

18. D. Ricci to G. Ricci, February 4, 1890, Fondo Ricci Curbastro, Biblioteca, Lugo.

19. U. Dini to P. Boselli, April 2, 1890, ACS, MPI, DGIS, UIS, busta 563, fasc. 797, sottofasc. 8.

20. L. Cremona, "Consiglio Superiore di Pubblica Istruzione, Estratto di verbale dell'adunanza," October 28, 1890, ACS, MPI, DGIS, UIS, busta 948, fasc. 797, sottofasc. 8.

21. T. Levi-Civita, "Commemorazione," 555.

22. Ibid., 561.

23. His academic studies through these years can be followed in Università degli Studi di Padova, Archivio Storico, Registro della carriera scolastica Tullio Levi-Civita.

24. G. Veronese, *Fondamenti di geometria a più dimensioni e a più specie di unità rettilinee esposti in forma elementare: Lezioni per la scuola di Magistero in matematica* (Padua, 1891).

25. Detlef Laugwitz, "Debates about Infinity in Mathematics around 1890: The Cantor–Veronese Controversy, its Origins and its Outcome," *NTM International Journal of History and Ethics of Natural Sciences, Technology & Medicine,* 10 (2002), 103; see also, D. Laugwitz, "Tullio Levi-Civita's Work on Nonarchimedean Structures," in *Levi-Civita: Convegno internationale celebrativo del centenario della nascita,* 8 (Rome, 1975), 297-311.

26. T. Levi-Civita, "Sugli infiniti ed infinitesimi attuali quali elementi analitici," *Atti Inst. R. Veneto di Sc., lett., ed arti,* 4(1892–1893), 1765–1815. The algebraic work is regarded today as a precursor to non-standard analysis.

27. U. Amaldi, "Commemorazione del Socio Tullio Levi-Civita," 1132.

28. T. Levi-Civita, "Sui numeri transfiniti," *Rend. Acc. dei Lincei,* 7 (1898), 91–96.

29. The following titles are listed in Registro della carriera scolastica Vol. D, no. 125, Università di Padova, Archivio Storico, Tullio Levi-Civita: "I numeri reali ammettono una legge generale di reciprocità", "Si può assegnare un criterio generale per la determinazione dell'ordine delle varietà generate da forme proiettive;" "Il metodo di Jacobi per la soluzione dei problemi di Dinamica può applicarsi al problema della funicolare, quando le forze hanno una funzione potenziale."

30. Lightly modified and published by Levi-Civita with the same title, "Sugli invarianti assoluti," in *Atti Inst. Veneto,* 5 (1893–94), 1447–1523.

31. T. Levi-Civita, "Commemorazione," 555.

32. ACS, MPI, DGIS, UIS, 1897–1910, busta 28.

33. Federigo Enriques to Guido Castelnuovo, November 27, 1894, quoted in Pietro Nastasi and Rossana Tazzioli, eds. "Tullio Levi-Civita," *Lettera matematica Pristem,* no. 57–58 (Milan, 2006).

34. Early in 1896, Levi-Civita had been appointed an internal professor in mathematics at the Scuola Normale in Scienze attached to the University of Pavia. He held that position until the end of the year.

35. Ernesto Pascal, Ferdinando Aschieri, Carlo Somigliana, "Relazione," May 2, 1896, ACS, MPI, DGIS, 1897–1910, busta 28.

36. F. D'Arcais, G. Ricci, V. Volterra, "Relazione," July 21, 1896, ibid. Translation by M. Vallisneri.

37. Eugenio Valli to unknown, November 22, 1896, ACS, MPI, DGIS, 1897-1910, busta 28.

38. Quoted in U. Bottazzini, A. Conte, and P. Gario, eds., *Riposte armonie* (Turin, 1996), footnote 348, which describes more fully this episode.

39. V. Volterra to Giovanni Battista Guccia, December 11, 1895, quoted in P. Nastasi, http://www.treccani.it/enciclopedia/tullio-levi-civita (2016).

40. Luca Gianturco to unknown, Jan. 7, 1898, Archivio Generale, Ateneo dell'Università degli Studi di Padova, fascicolo- docente Tullio Levi-Civita. Apparently, Marcolongo was a closet anti-Semite and nursed a long-standing grudge against Levi-Civita and other Jewish university professors in the mathematical community, which became more pronounced after the passage of Italy's racial laws in 1938. For the details, see Giorgio Israel, *Il fascismo e la razza. La scienza italiana e le politiche razziali del regime* (Bologna, 2010), chapter 6.

41. V. Volterra, "Premio Lincei Matematica, 1909," Vito Volterra Papers, Accademia Nazionale dei Lincei, Rome.

42. T. Levi-Civita, "Sulle trasformazioni delle equazioni dinamiche," *Ann. di Mat.*, 24 (1896), 227.

43. U. Amaldi, "Tullio Levi-Civita," *Opere matematiche,* vol. 1, xiii.

44. G. Ricci to T. Levi-Civita, July 27, 1918, Tullio Levi-Civita Papers, Accademia Nazionale dei Lincei, Rome.

## Notes for Chapter 9

1. G. Ricci to B. Ricci, October 26, 1894, Fondo Ricci Curbastro, Lugo.

2. Ibid., October 31, 1894.

3. Ibid., November 3, 1894.

4. "It has been my experience that beneath the surface of every scientist there lurks a wounded person who never feels his work has been fully appreciated. Feynman was a rare, perhaps even unique, exception." David Goodstein, "Richard Feynman: In Memoriam," *California Tech,* February 19, 1988, p. 4.

5. T. Levi-Civita, "Commemorazione," 558.

6. The prize was offered "to study the surfaces whose linear element can be reduced to the form: $ds^2 = [f(u)\phi(v)](du^2 + dv^2)$," *Bull. Amer. Math. Soc.,* 2 (1893), 260–261.

7. E. Beltrami to G. Ricci, October 3, 1892, Fondo Ricci Curbastro, Lugo.

8. Gabriel Koenigs, *Notice sur les travaux scientifiques de Gabriel Koenigs* (Tours, 1897), 26. The following year, the French Academy gave him the Poncelet Prize for a group of works in geometry and mechanics.

9. G. Koenigs, "Résumé d'un mémoire sur les lignes géodésiques," *Ann. Fac. Sc. de Toulouse* 6 (1892), 1-34.

10. G. Ricci, "Dei sistemi di coordinate atti a ridurre la espressione del quadrato dell'elemento lineare di una superficie alla forma $ds^2 = (U+V)(du^2+dv^2)$, *Rend. Acc. Lincei* 5 (1893), 73–81.

11. Jeremy Gray, "A History of Prizes in Mathematics," in J. Carlson, A. Jaffe, and A. Wiles, eds., *The Millennium Prize Problems* (Providence, 2006), 16.

12. G. Ricci, "A proposito di una memoria sulle linee geodetiche del Sig. G. Koenigs," *Rend. Acc. Lincei,* 5 (1893), 146–148.

13. G. Koenigs, "Réponse à la note de Monsieur le professeur Gregorio Ricci, du 3 septembre 1893," *Rend. Acc. Lincei,* 5 (1893), 336.

14. G. Ricci, "Alcune parole a proposito della precedente risposta del sig. Koenigs," *Rend. Acc. Lincei* 2 (1893), 353. An excerpt of Koenigs's "Réponse à la note de M. le professeur Gregorio Ricci, du 3 septembre 1893," *Rend. Acc. Lincei,* 2 (1893), 336–337, is reproduced as a footnote in Ricci, *Opere,* vol. 1, 351.

15. G. Koenigs, "Résumé," *Ann. Fac. Sc. de Toulouse* 6(1892).

16. G. Ricci, "Sulla teoria delle linee geodetiche e dei sistemi isotermi di Liouville," *Atti Ist. Veneto* 5 (1894), 356.

17. G. Koenigs, "Mémoire sur les lignes géodésiques," Mém. *Savants Étrang.* 6 (1894), 1–318.

18. Richard Dedekind, a student of Gauss, was a German mathematician who made important contributions to the theory of the real number system in terms of arithmetic properties of the rational number system today known as "Dedekind cuts."

19. Undated manuscript of T. Levi-Civita, III Raccoglitore, "Disputa Capelli-Ricci," Fondo Ricci Curbastro, Lugo.

20. G. Ricci, "In memoria di Alfredo Capelli," *Opere,* vol. 2, 331.

21. See Chapter 8, page 66 for a fuller discussion of this paper.

22. T. Levi-Civita to V. Volterra, April 8, 1896, in Pietro Nastasi and Rosanna Tazzioli, eds., *Aspetti scientifici e umani nella corrispondenza di Tullio Levi-Civita* (Milan, 2000), 22.

23. G. Ricci to T. Levi-Civita, July 3, 1896, Tullio Levi-Civita Papers, Accademia Nazionale dei Lincei, Rome.

24. G. Koenigs, "Mémoire sur les lignes géodésiques," Mém. *Savants Étrang.*, No. 631 (1894), 318 pp.

25. G. Ricci to T. Levi-Civita, July 7, 1896, "La formazione," in P. Nastasi and R. Tazzioli, eds., *Tullio Levi-Civita, Lettera matematica pristem* 57/58 (Milan, 2006), 13.

26. T. Levi-Civita to G. Ricci, July 9, 1896, ibid.

27. T. Levi-Civita, "Sulle transformazioni," 241.

28. T. Levi-Civita to G. Ricci, July 31, 1896, Fondo Ricci Curbastro, Lugo.

29. T. Levi-Civita, *Opere,* vol. 1, 5.

30. T. Levi-Civita to Arnold Sommerfeld, March 30, 1899, Sommerfeld Papers, Deutsches Museum (Archiv HS 1977-28/A, 200).

31. T. Levi-Civita to V. Volterra, November 28, 1899, in Nastasi and Tazzioli, *"Aspetti scientifici,"* 57.

32. G. Ricci and T. Levi-Civita, "Méthodes de calcul différentiel absolu et leurs applications," Math. Ann. 54 (1900), 125–201.

33. "Ricci and Levi-Civita's Tensor Analysis Paper," in Robert Hermann, trans. and ed., *Lie Groups: History, Frontiers, and Applications,* vol. 2 (Brookline, 1975), iii. Hermann's translation is a modern rendition of the original article, rather than a literal translation. The volume uses modern notation and terminology, starting with the substitution of the term "tensor analysis" for "absolute differential calculus." The text also contains additional material, including a number of "Remarks" designed to bring the material into line with contemporary differential geometry and physics.

34. Quoted in G. Ricci and T. Levi-Civita, "Méthodes," 128. Translation by D. Babbitt.

35. G. Ricci, *Lezioni sulla teoria delle superficie* (Padua, 1898); G. Ricci, *Lezioni sulla teoria matematica dell' elasticità,* completed in 1901 and published posthumously in 1957, in Ricci, *Opere,* vol. 2.

36. The Drucker brothers, Jewish booksellers who catered to the university community, ran their lively business from quarters in the historic palazzo del Bò.

37. George Oscar James, Review. *Bull. Amer. Math. Soc.* 7(1901), 359–360.

38. G. Ricci to V. Volterra, July 5, 1888, Vito Volterra Papers, Accademia dei Lincei.

39. G. Castelnuovo, "Commemorazione del socio Federigo Enriques," *Rend. Acc. Naz. Lincei,* 8 (1947), 3–21. See also D. Babbitt and J. Goodstein, "Guido Castelnuovo and Francesco Severi: Two Personalities, Two Letters," *Notices of the AMS,* 56 (2009), 800–801.

40. U. Bottazzini, "Ricci and Levi-Civita," 248. See also, U. Bottazzini, A. Conte, and P. Gario, "La relazione di Castelnuovo ed Enriques. Documenti inediti per il premio Reale di Matematica del 1901," *Studies in the History of Modern Mathematics,* III, *Suppl. Rendiconto del Circolo Matematico di Palermo,* 55 (1998), 75-156.

41. U. Bottazzini et al, "La relazione," 127.

42. Ibid.

43. F. Enriques to G. Castelnuovo, April 18, 1903, in U. Bottazzini, A. Conte, and P. Gario, eds., *Riposte armonie. Lettere di Federigo Enriques a Guido Castelnuovo* (Turin, 1996), 559.

44. L. Bianchi, "Relazione sul concorso al Premio Reale per la Matematica per l'anno 1901," *Atti Acc. Lincei Rend. Sedute solenni,* 2 (1902–1913), 142–151, on p. 147.

45. L. Bianchi, "Relazione," 148.

46. Ibid., 149; G. Ricci, "Dei sistemi di congruenze ortogonali in una varietà qualunque," *Opere,* vol. 2, 1–61.

47. L. Bianchi, "Relazione," 150.

48. Ibid.

## Notes for Chapter 10

1. Louis Kollros, "Albert Einstein en Suisse. Souvenirs," *Helv. Phys. Acta Suppl.* 4 (1956), 278.

2. Abraham Pais, *Subtle Is the Lord: The Science and the Life of Albert Einstein* (New York, 1982), 201. See also the editorial note "Einstein on Gravitation and Relativity: The Static field," *The Collected Papers of Albert Einstein* (CPAE hereafter), Vols. 1–14 (Princeton, 1987–).

3. CPAE, vol. 6, Doc. 42, p. 418.

4. In 1915, Einstein predicted twice that amount of bending, based on his newly formulated theory of general relativity.

5. Albert Einstein to Willem Julius, Sept. 22, 1911, in CPAE, vol. 5, Doc. 288, p.209.

6. In summer 1914, a solar-eclipse expedition led by the young German astronomer Erwin Finlay-Freundlich set out for the Crimea, only to be caught up in the outbreak of World War I before it could test Einstein's prediction. Finlay-Freundlich was captured by Russian soldiers but released later in a prisoner exchange. The British full-solar-eclipse expeditions of 1919 provided an exceptional viewing opportunity since the eclipse positioned the sun in front of the Hyades, a particularly bright cluster of stars. For further detail, see Jeffrey Crelinsten, *Einstein's Jury: The Race to Test Relativity* (Princeton, 2006).

7. See Stefanie U. Eminger, "Carl Friedrich Geiser and Ferdinand Rudio: The Men Behind the First International Congress of Mathematicians," (Ph.D. diss., University of St. Andrew, 2015), 129, http://hdl.handle.net/10023/6536; A. Einstein, *Autobiographical Notes,* ed. and trans. Paul Arthur Schilpp (La Salle, 1979), 17; Anton Reiser, *Albert Einstein: A Biographical Portrait* (New York, 1930), 49.

8. Carl Seelig, *Albert Einstein* (London, 1956), 28. My thanks to Dennis Lehmkuhl for calling my attention to this quotation.

9. Peter L. Galison, "Minkowski's Space-Time: From Visual Thinking to the Absolute World," *Hist. Studies in the Phys. Sci.* 10 (1979), 96.

10. Hermann Minkowski, "Raum und Zeit," 1908, trans. D. Lehmkuhl, in *Minkowski Spacetime: A Hundred Years Later,* ed. Vesselin Petkov (Springer, 2010), xv–xlii, on xv.

11. Scott Walter, "Minkowski's Modern World," in V. Petkov, *Minkowski Spacetime,* 43–61, on 56; see also, S. Walter, "Minkowski, Mathematicians, and the Mathematical Theory of Relativity," in H. Goenner, J. Renn, J. Ritter, T. Sauer, eds. *The Expanding Worlds of General Relativity,* Einstein Studies, vol. 7 (1999), 45–86.

12. S. Walter, "Minkowski," 59.

13. Jean Eisenstaedt, *The Curious History of Relativity: How Einstein's Theory of Gravity Was Lost and Found Again* (Princeton, 2006), 44.

14. Anton Reiser, *Albert Einstein* (New York, 1930), 49.

15. A. Einstein, *Autobiographical Notes,* ed. and trans. Paul Arthur Schilpp (La Salle, 1996), 15.

16. Jakob Laub to A. Einstein, CPAE, vol. 5, Doc. 79, Feb. 2, 1908.

17. A. Pais, *Subtle Is the Lord,* 152. Pais is referring to the tensors of Gibbs and Heaviside, which are useful in physics, but have no relation to the curved manifolds addressed by the absolute differential calculus.

18. CPAE, vol. 2, Doc. 51, p. 329; see also the editorial note, "Einstein and Laub on the electrodynamics of moving media," vol. 2, pp. 503-507.

19. A. Einstein to Mileva Einstein-Marić, April 17, 1908, CPAE, vol. 5, Doc. 96, p. 69.

20. J. Laub to A. Einstein, May 18, 1908, CPAE, vol. 5, Doc. 101, pp. 72–73, on 73.

21. A. Einstein to J. Laub, July 30, 1908, CPAE, vol. 5, Doc. 113, pp. 81–82, on 82.

22. A. Einstein, "Fundamental Ideas and Methods of Relativity Theory, Presented in Their Development," [after 22 Jan. 1920], unpublished draft of an article for *Nature,* CPAE, vol. 7, Doc. 31.

23. A. Einstein, "How I Created the Theory of Relativity," trans. Y. A. Ono, *Physics Today,* 35(1982), 45–47, on 47. A new, definitive translation appears in CPAE, vol. 13, Doc. 399, see p. 638.

24. Philipp Frank, *Einstein: His Life and Times,* trans. George Rosen (New York, 1947), 82.

25. CPAE, vol. 4, Doc. 8, p. 133.

26. Quoted in Tilman Sauer, "Marcel Grossmann and his contribution to the general theory of relativity," p. 3, Apr. 22, 2014, arXiv:1312.4068v2 [physics. hist-ph].

27. Ibid., 42.

28. More about Grossmann's early life is related in the recent biography by his granddaughter, Claudia E. Graf-Grossmann, *Aus Liebe zur Mathematik* (Zurich, 2015). An English-language edition is not currently available.

29. Banesh Hoffmann (with Helen Dukas), *Albert Einstein, Creator and Rebel,* (New York, 1972), 25.

30. Maja Winteler-Einstein, "Albert Einstein—A Biographical Sketch," CPAE, vol. 1, xxi.

31. Hermann Einstein to Jost Winteler, Dec. 30, 1895, CPAE, vol. 1, Doc. 14.

32. T. Sauer, "Marcel Grossmann," 4.

33. C. Seelig, Albert Einstein, 34.

34. Ibid., 207-208.

35. P. Frank, *Einstein,* 102.

36. L. Kollros, "Albert Einstein," 278.

37. T. Sauer, "Marcel Grossmann." 42.

38. A. Pais, *Subtle Is the Lord,* 213.

39. Ibid.

40. T. Sauer, "Marcel Grossmann." 15.

41. Quoted in T. Sauer, ibid., 16.

42. CPAE, vol. 6, Doc. 42, p. 418.

43. Quoted in S. U. Eminger, "Carl Friedrich Geiser," 21.

44. CPAE, vol. 13, Doc. 399, p. 638.

45. While he never got around to publishing these lectures, they are included in volume 2 of Ricci's *Opere,* published in 1957.

46. A. Einstein to A. Sommerfeld, October 29, 1912, quoted in J. Goodstein, "The Italian Mathematicians of Relativity," *Centaurus,* 26 (1983), 247.

47. Quoted in A. Pais, *Subtle Is the Lord,* 212.

48. A. Einstein and Marcel Grossmann, *Entwurf einer verallgemeinerten Relativiäts-theorie und einer Theorie der Gravitation* (Leipzig, 1913); CPAE, vol.4, Doc. 13.

49. Quoted in T. Sauer, "Marcel Grossmann," 44.

50. Ibid., 18–19.

51. A. Einstein, "Die Grundlage der allgemeinen Relativitätstheorie," *Annalen der Physik,* 49 (1916), 769-822 ["The Foundation of the General Theory of Relativity"].

52. T. Levi-Civita, "Gregorio Ricci-Curbastro," *Opere,* vol. 1, 4.

53. Ibid., 11.

54. J. Eisenstaedt, *The Curious History of Relativity,* 60.

55. These gravitational potentials are the components $g_{ij}$ of the Riemannian metric tensor. In other words, the right-hand side represents non-gravitational physics; the left-hand side is the geometrical representation of gravity.

56. James Hartle, *Gravity: An Introduction to Einstein's General Relativity* (Boston, 2003), 13.

57. Quoted in Walter Isaacson, *Einstein,* 196.

## Notes for Chapter 11

1. T. Levi-Civita to Max Abraham, Feb. 7, 1907, in "1908 Congress (IV) of Mathematicians, Letters," Levi-Civita Family Papers.

2. Max Born and Max von Laue, "Max Abraham," *Physikalische Zeitschrift,* 24(1923), 53; see also Claudio Citrini, "Matematica e vita civile nel Politecnico di cento anni fa: la vicende di Max Abraham," *Annali di Storia delle Università Italiane,* 12 (2008), 104.

3. The first volume (Leipzig, 1904–1905) was based on an earlier work by August Föppl on Maxwell's theory of electricity (1894); later editions appeared in 1908, 1912, 1914, and 1918.

4. M. Abraham to T. Levi-Civita, Feb. 15, 1907, "1908 Congress," Levi-Civita Family Papers.

5. For further detail, see Stanley Goldberg, *Understanding Relativity* (Boston, 1984), 103–149.

6. V. Volterra, "Le Matematiche in Italia nella seconda metà del secolo XIX," in G. Castelnuovo, ed., *Atti del IV Congresso Internazionale dei Matematici,* vol. 1 (Rome, 1909), 64.

7. For further details of Abraham's stay in America, see Lawrence Badash, ed., *Rutherford and Boltwood, Letters on Radioactivity* (New Haven, 1969), 193–194.

8. T. Levi-Civita to unknown recipient, draft, ca. 1909, Fondo Levi-Civita, Accademia Nazionale dei Lincei, Rome.

9. File, "Relazione della Commissione Giudicatrice, Concorso per Professore Straordinario di Meccanica Razionale nel R. Istituto Tecnico Superiore di Milano," Oct. 8, 1909, Levi-Civita Family Papers.

10. A. Pais, *Subtle Is the Lord,* 156.

11. Quoted in A. Pais, *Subtle Is the Lord,* 232.

12. See Jagdish Mehra, "Einstein, Hilbert, and the Theory of Gravitation," in J. Mehra, ed., *The Physicist's Conception of Nature* (Dordrecht, 1973), 96–98.

13. A. Einstein to Lucien Chavan, ca. July 6, 1911, CPAE, vol. 5, Doc. 271, pp. 193–194.

14. A. Einstein to Heinrich Zangger, [before Feb. 29], 1912, CPAE, vol. 5, Doc. 366, p. 268.

15. A. Einstein to J. Laub, Aug. 11, 1911, CPAE, vol. 5, Doc. 275, p. 197.

16. For the best recent study of Abraham's theory of gravitation, see Jürgen Renn, "The Summit Almost Scaled: Max Abraham as a Pioneer of a Relativistic Theory of Gravitation," in J. Renn, ed., *The Genesis of General Relativity,* vol. 3 (Dordrecht, 2007), 1229-1255.

17. A. Einstein to Michele Besso, Mar. 26, 1912, CPAE, vol. 5, Doc. 377, pp. 276–279, on 278.

18. A. Einstein to H. Zangger, Jan. 27, 1915, vol. 5, Doc. 344, p. 250.

19. A. Einstein to Paul Ehrenfest, Feb. 12, 1912, vol. 5, Doc. 357, p. 260.

20. A. Einstein, "The Speed of Light and the Statics of the Gravitational Field," *Annalen der Physik,* CPAE, vol. 4, Doc 3, p. 95. In March, 1912, Einstein published his final paper on the static field, "On the Theory of the Static Gravitational Field," CPAE, vol. 4, Doc. 4.

21. Carlo Cattani and Michelangelo De Maria, "Max Abraham and the Reception of Relativity in Italy: His 1912 and 1914 Controversies with Einstein," in Don Howard and John Stachel, eds., *Einstein and the History of General Relativity* (Boston, 1896), 160-174, on 163.

22. A. Einstein, "Relativity and Gravitation: Reply to a Comment by M. Abraham," *Annalen der Physik* 38(1912), 1059–1064; CPAE, vol. 4, Doc. 8. For a more detailed account, see the editorial note, "Einstein on Gravitation and Relativity: The Static Field," vol. 4, pp. 122–128.

23. CPAE, vol. 4, Doc. 8, p. 133.

24. Quoted in C. Cattani and M. De Maria, "Abraham and Relativity," 164.

25. A. Einstein, "Comment on Abraham's Preceding Discussion 'Once Again, Relativity and Gravitation,' " *Annalen der Physik* 39 (1912), 704; CPAE, vol. 4, Doc. 9.

26. T. Sauer to the author, Apr. 27, 2017.

27. Quoted in C. Cattani and M. De Maria, "Abraham and Relativity," 165.

28. A. Einstein to M. Besso, Mar. 12, 1912, CPAE, vol. 5, Doc. 377.

29. A. Einstein to A. Sommerfeld, Oct. 29, 1912, vol. 5, Doc. 421.

30. A. Einstein to Alfred Kleiner, April 3, 1912, vol. 5, Doc. 382.

31. T. Levi-Civita to Giuseppe Colombo, Mar. 17, 1913, quoted in C. Citrini, "Matematica e Vita Civile," 103.

32. Ibid., 102.

33. Quoted in M. Born and M. von Laue, "Max Abraham," *Phys. Zs.,* 24 (1923), 53.

34. A. Einstein to M. Besso, [after Jan. 1, 1914], Mar. 26, 1912, CPAE, vol. 5, Doc. 499.

35. Quoted in Barbara J. Reeves, "Einstein Politicized: The Early Reception of Relativity in Italy," in Thomas F. Glick, ed., *The Comparative Reception of Relativity* (Dordrecht, 1987), 201.

36. F.G. Tricomi, "Mathematici italiani del primo secolo dello stato unitario," *Mem. Acc. Sci. Torino,* cl. sci., fis., 4th ser. 1(1962), 106.

37. C. Somigliana, *Tullio Levi-Civita e Vito Volterra* (Milan, 1946), 8.

38. A. Einstein, "The Formal Foundation of the General Theory of Relativity," *Königlich Preussische Akademie der Wissenschaften* (Berlin). *Sitzungsberichte,* CPAE, vol. 6, Doc. 9, p. 30. Unlike the title of the two previous *Entwurf* papers, Einstein replaced the word "generalized" with "general" in describing his theory of relativity.

39. Hanoch Gutfreund and Jürgen Renn, *The Road to Relativity*: The History and Meaning of Einstein's "The Foundation of General Relativity," (Princeton, 2015), 26.

40. A. Einstein, "The Formal Foundation," CPAE, vol. 6, Doc. 9.

41. M. Abraham to T. Levi-Civita, Feb. 23, 1915, Tullio Levi-Civita Papers, Lincei.

42. A. Pais, *Subtle Is the Lord,* 244.

43. C. Somigliana, "Tullio Levi-Civita," 6–7.

44. For extensive analysis, see C. Cattani and M. De Maria, "The 1915 Epistolary Controversy between Einstein and Tullio Levi-Civita," in Howard and Stachel, eds., *Einstein and the History of General Relativity,* vol. 1, 175–200, on 194.

45. John Norton, "How Einstein Found His Field Equations, 1912–1915," ibid., 101–159, on 138.

46. H. Lorentz, Jan. 1, 1916, CPAE, vol. 8, Doc. 177.

47. A. Einstein to T. Levi-Civita, Mar. 5, 1915, CPAE, vol. 8, Doc. 60, p. 71. Portions of the material on Einstein and Levi-Civita previously appeared in the journal *Centaurus* and are reprinted here in slightly different form with the consent of John Wiley & Sons.

48. T. Sauer, "Albert Einstein, Review Paper on General Relativity Theory (1916)," I. Grattan-Guinness, ed., *Landmark Writings in Western Mathematics, 1640–1940,* chap. 63 (Amsterdam, 2005), 811.

49. A. Einstein to T. Levi-Civita, Mar. 17, 1915, CPAE, vol., 8, Doc. 62, p. 73.

50. A. Einstein to T. Levi-Civita, Mar. 26, 1915, CPAE, vol. 8, Doc. 66.

51. M. Abraham to T. Levi-Civita, Mar. 30, 1915, Tullio Levi-Civita Papers, Lincei.

52. A. Einstein to T. Levi-Civita, Apr. 2, 1915, CPAE, vol. 8, Doc. 69.

53. A. Einstein to H. Zangger, April 10, 1915, CPAE, vol. 8, Doc. 73.

54. A. Einstein to T. Levi-Civita, Apr. 14, 1915, CPAE, vol 8, Doc., 75.

55. For further details, see C. Citrini, "Matematica e Vita Civile," 109–115.

56. M. Abraham to T. Levi-Civita, Apr. 27, 1915, Tullio Levi-Civita Papers, Lincei.

57. Ibid., Oct. 16, 1919.

58. See M. Abraham's letter to Theodore von Kármán, Feb. 16, 1922, Theodore von Kármán Papers, Institute Archives, California Institute of Technology, Box 1.1.

59. T. von Kármán to A. Einstein, Feb. 22, 1922, CPAE, vol. 13, Doc. 61 and "Auszüge Aus Den Gutachten Über Max Abraham," undated, but after Feb. 22, in T. von Kármán Papers, Abraham file, Box 1.1.

60. W. Isaacson, *Einstein,* 214.

61. A. Einstein to A. Sommerfeld, Nov. 28, 1915, CPAE, vol. 8, Doc. 153.

62. A. Einstein, "On the General Theory of Relativity," Nov. 4, 1915, CPAE, vol. 6, Doc. 21.

63. A. Einstein to P. Ehrenfest, Jan. 17, 1916, CPAE, vol. 8, Doc. 182.

64. A. Pais, *Subtle Is the Lord,* 253.

65. Ibid.

66. J. Norton, "How Einstein Found His Field Equations," 148.

67. Ibid., 142.

68. A. Einstein to A. Sommerfeld, Nov. 28, 1915, CPAE, vol. 8, Doc. 153.

69. A. Einstein, "On the General Theory" Nov. 4, 1915, CPAE, vol. 6, Doc. 21.

70. A. Einstein, "The Field Equations of Gravitation," Nov. 25, 1915, CPAE, vol. 6, Doc. 25.

71. A. Einstein, "The Foundation of the General Theory of Relativity," Mar. 20, 1916, CPAE, vol. 6, Doc. 30.

72. Ibid., p. 153.

## Notes for Chapter 12

1. Libera and Tullio Levi-Civita to Trevisani family, August 2, 1914, Levi-Civita family correspondence.

2. Claudio G. Segrè, Italo Balbo: *A Fascist Life* (Berkeley, 1987), 18.

3. R. J. B. Bosworth, *Mussolini's Italy: Life under the Fascist Dictatorship,* 1915–1945 (New York, 2007), 58.
4. S.S.C. Interview.
5. Ibid.
6. Ibid.
7. Ibid.
8. Ibid.
9. Ibid.
10. See for example, U. Bottazzini, "Ricci and Levi-Civita," 254.
11. T. Levi-Civita to George Birkhoff, Jan. 9, 1917, quoted in J. Goodstein, "The Italian Mathematicians of Relativity," 255.
12. T. Levi-Civita, "Nozione di parallelism in una varietà qualunque e conseguente specificazione geometrica della curvatura Riemanniana," *Rend. Circ. Mat. di Palermo* 42 (1917), 173–215; T. Levi-Civita, *Opere matematiche,* vol. 4 (Rome, 1960), 1–39.
13. Ibid., 197.
14. Ibid.
15. G. Castelnuovo to T. Levi-Civita, April 19, 1917, Tullio Levi-Civita Papers, Accademia Nazionale dei Lincei, Rome. For Severi's approach to the notion of parallelism, see F. Severi, "Sulla curvatura delle superficie e varietà," *Rend. Circolo Mat. Palermo* 42 (1917), 227–259.
16. U. Bottazzini, "Ricci and Levi-Civita," 254.
17. O. Darrigol, "The Mystery of Riemann's Curvature," 77.
18. George Y. Rainich, "Levi-Civita on Tensor Calculus," *Bull. Amer. Math. Soc.* 34 (1928), 775–776. For more on this fundamental contribution of Levi-Civita to the advance in the absolute differential calculus, see W. V. D. Hodge, "Tullio Levi-Civita, 1873–1941," Appendix C.
19. In simpler, modern terms, one speaks of a curved space embedded in a larger-dimensional Euclidian (i.e., flat) space. To move vectors along the curved surface, one follows their motion in the embedding space while projecting them "down" to the surface. Nevertheless, a manifold's Riemann curvature prescribes the displacement of vectors even when the manifold is not embeddable.
20. Harold S. Ruse, "Tullio Levi-Civita," *Edinburgh Math. Notes* 33 (1943), 21.
21. D. Struik, "Schouten, Levi-Civita, and the Emergence of Tensor Calculus," in David Rowe and John McCleary, eds., *The History of Modern Mathematics*, vol. 2 (Boston, 1989), 103.
22. Ibid., 103–104.

23. Jan A. Schouten, *Der Ricci-Kalkül* (Berlin, 1924); J.A. Schouten, *Ricci-Calculus: An Introduction to Tensor Analysis and its Geometrical Applications* (Berlin, 1954).

24. Enrico Sangiorgi to the author, Mar. 24, 2014; Armanda Capucci, "La vecchia casa di Gregorio Ricci Curbastro," [Internet], retrieved June 23, 2014, available from http://www/pavaglionelugo.net/2014/06/la-vecchia-casa-di-gregorio-ricci.html.

25. G. Ricci to Giorgio Ricci Curbastro, Nov. 26, 1915, Fondo Ricci Curbastro, Lugo.

26. G. Ricci to Cesare Ricci Curbastro, Feb. 24, 1916, Fondo Ricci Curbastro, Lugo.

27. Cesare later met and married Giulia Archi.

28. G. Ricci to Giorgio Ricci Curbastro, September 10, 1917, Fondo Ricci Curbastro, Lugo.

29. S.S.C. Interview.

30. G. Castelnuovo to T.Levi-Civita, Aug. 1, 1918, in the Levi-Civita Papers, Rome; ACS, MPI, DGIU, fasc. professori universitaria, III serie, busta 267.

31. T. Levi-Civita, "Come potrebbe un conservatore giungere alla soglia della nuova meccanica," *Opere matematiche,* vol. 4, 197.

32. *Il Messaggero,* Oct. 30, 1921.

33. G. Ricci to T. Levi-Civita, Oct. 31, 1921, Levi-Civita Papers, Rome.

34. "Il Prof. Einstein a Padova," *Corriere della Sera,* October 28, 1921.

35. F. Toscano, *Il genio,* 253–255.

36. F. Tricomi, *La mia vita di matematico attraverso la cronistoria dei miei lavori* (Padova, 1967), 7.

37. Fragment of Ricci's inscription on Elio's gravestone, quoted in Armanda Capucci, "La vecchia casa di Gregorio Ricci Curbastro," http://www.pavaglionelugo.net/2014/06/la-vecchia-casa-di-gregorio-ricci.html [retrieved September 29, 2014].

38. T. Levi-Civita, "Commemorazione," 562.

39. G. Ricci, "Direzioni e invarianti principali in una varietà qualunque," *Opere,* vol. 2, 315.

40. Marcel Berger, *A Panoramic View of Riemannian Geometry* (Berlin, 2003), 245.

41. Applications of Ricci's curvature in Riemannian geometry mostly stayed under the mathematical world's radar until 1940, when S. B. Myers, "Riemannian Manifolds with Positive Mean Curvature," *Duke Math. Journal* 8(1941), 401–404, obtained a significant new result in Riemannian geometry using Ricci's curvature tensor.

42. T. Levi-Civita, "Sulla espressione analitica spettante al tensore gravitazionale nella teoria di Einstein," in *Opere matematiche,* vol. 4, 63.

43. T. Levi-Civita, *The Absolute Differential Calculus,* ed. Enrico Persico and transl. Marjorie Long (New York: Dover Publications, 1977), 200.

44. Enrico Bompiani, "La géométrie des espaces courbes et le tenseur d'énergie d'Einstein," *Comptes rendus* 174 (1922), 737. I thank Tilman Sauer for this reference.

45. Quoted in Solomon Lefschetz, "Luther Pfahler Eisenhart, 1876–1965," *Biographical Memoirs of the National Academy of Sciences* 40 (1969), 71.

46. H. S. Ruse, "The Ricci Calculus, *Nature* 171 (1953), 62.

47. Luther P. Eisenhart to Oswald Veblen, May 7, 1912, Oswald Veblen Papers, Manuscript Division, Library of Congress, Washington, D.C.

48. Leopold Infeld, *Quest: The Evolution of a Scientist* (London, 1941), 238.

49. L. P. Eisenhart, "Ricci's Principal Directions for a Riemann Space and the Einstein Theory," *Proc. Nat. Acad. Sci.* 8 (1922), 19.

50. L. P. Eisenhart, *Riemannian Geometry* (Princeton, 1926), iii.

51. D. Struik, "J.A. Schouten and the tensor calculus," *Nieuw Archief Voor Wiskunde* 26 (1978), 103.

52. G. Ricci, "Sulle varietà a invarianti principali equali," *Rend. Acc. Lincei* 33 (1924), 431.

53. J. Eisenstaedt, *The Curious History of Relativity,* 131.

54. The Einstein equations (or equivalently, the action, written in terms of the Ricci scalar, the contraction of the Ricci tensor) are also subsumed in string theory, quantum gravity, and alternative theories of gravity.

55. Dana Mackenzie, "The Poincaré Conjecture—Proved," *Science* 314(2006), 1848–1849.

56. G. Ricci, "Sulle varietà," 432.

57. For an introduction to the historical issues, see David Rowe, "Einstein's Gravitational Field equations and the Bianchi Identities," *Mathematical Intelligencer* 24(2002), 57–66.

58. F. Toscano, "Luigi Bianchi, Gregorio Ricci Curbastro e la scoperta delle identità di Bianchi," Atti del XX congresso nazionale di storia della fisica e dell'astronomia, CUEN, Napoli (2001), 361.

59. T. Levi-Civita, *The Absolute Differential Calculus,* ed. Enrico Persico (New York, Dover Publications, 1977), 182.

60. T. Levi-Civita, "Commemorazione," 564.

61. H. S. Ruse, "The Ricci Calculus," *Nature* 171 (1953), 61–62.

62. F. G. Tricomi, Remarks, in "Tullio Levi-Civita, Convegno internazionale celebrativo del centenario della nascita," Atti dei Convegni Lincei 8 (1975), 18–19.

63. Benito Mussolini, quoted in R. J. Bosworth, *Mussolini's Italy,* 240.

64. For an account of Volterra's lonely defiance of Mussolini's regime, see J. Goodstein, *The Volterra Chronicles: The Life and Times of an Extraordinary Mathematician* (Providence, 2007), 193–200.

65. Giuseppe Levi to T. Levi-Civita, November 8, 1931, quoted in P. Nastasi, "La comunità matematica italiana di fronte alle legge razziali," in Massimo Galuzzi, ed., *Giornate di storia della matematica: Cetraro, 1988* (Rende, 1991), 438.

66. T. Levi-Civita to Pietro de Francisci, November 19, 1931, "Giuramento, Archivio," Levi-Civita family papers.

67. W. V. D. Hodge, "Tullio Levi-Civita." See Appendix C.

68. J. Goodstein and D. Babbitt, "A Fresh Look at Francesco Severi," *Notices Math. Soc.* 59 (2012), 1070. Neugebauer subsequently resigned from all editorial duties at Springer journals, left Germany, and joined the Brown university faculty in 1939, where he became the founding editor of the AMS's new abstracting journal, *MR (Mathematical Reviews)*.

69. S.S.C. Interview.

70. C. Somigliana, "Tullio Levi-Civita e Vito Volterra," *Rend. Seminario matematico e fisico di Milano* 17 (1946), 2.

71. D. Struik to the author, November 6, 1973.

## Notes for Appendix A

1. E. P. Wigner. The unreasonable effectiveness of mathematics in the natural sciences. *Communications on Pure and Applied Mathematics*, 13(1): 1–14, 1960.

2. A. Einstein and M. Grossmann. *Outline of a generalized theory of relativity and of a theory of gravitation.* Teubner, Leipzig, 1913.

3. Einstein on gravitation and relativity: the collaboration with Marcel Grossmann. In M. J. Klein, A. J. Kox, J. Renn, and R. Schulman, editors, *The Collected Papers of Albert Einstein*, volume 4, page 294. Princeton University Press, Princeton, NJ, 1995.

4. A. Einstein. The field equations of gravitation. *Sitzungsberichte Königlich Preußische Akademie der Wissenschaften (Berlin)*, page 844, 1915.

5. K. Reich. Differential geometry. In I. Grattan-Guinness, editor, *Companion Encyclopedia of the History and Philosophy of the Mathematical Sciences*, volume 2, page 331. Johns Hopkins, Baltimore and London, 2003.

6. A. Zee. *Einstein gravity in a nutshell.* Princeton University Press, Princeton, NJ, 2013.

7. R. Cooke. *It's about time: elementary mathematical aspects of relativity.* American Mathematical Society, Providence, RI, 2017.

8. J. Renn, editor. *The Genesis of General Relativity.* Springer, Dordrecht, The Netherlands, 2007.

9. Einstein's research notes on a generalized theory of relativity. In M. J. Klein, A. J. Kox, J. Renn, and R. Schulman, editors, *The Collected Papers of Albert Einstein*, volume 4, page 192. Princeton University Press, Princeton, NJ, 1995.

10. G. Ricci and T. Levi-Civita. Méthodes de calcul différentiel absolu et leurs applications. *Mathematische Annalen*, 54:125–201, 1900.

11. A. Einstein. On the general theory of relativity. *Sitzungsberichte Königlich Preußische Akademie der Wissenschaften (Berlin)*, page 778, 1915; A. Einstein. Explanation of the perihelion motion of mercury from the general theory of relativity. *Sitzungsberichte Königlich Preußische Akademie der Wissenschaften (Berlin)*, page 831, 1915; A. Einstein. The field equations of gravitation. *Sitzungsberichte Königlich Preußische Akademie der Wissenschaften (Berlin)*, page 844, 1915.

12. One equation that uses the scalar product is the law of motion for a particle in a central potential,

$$m\frac{d^2x^i}{dt^2} = -\frac{\partial}{\partial x^i}U(r),$$

with $r^2 = \sum_i (x^i)^2$.

13. These forces are termed *fictitious* because they do not arise from physical interactions between bodies, but rather from describing physics from a non-inertial coordinate frame—in this case a rotating frame. Since these forces originate from geometry (more precisely, from kinematics), they affect bodies of different mass in the same way, as can be seen by dividing both sides of Eq. (7) by $m$.

14. A. Einstein. On the electrodynamics of moving bodies. *Annalen der Physik*, 17:891, 1905.

15. H. Minkowski. Space and time. *Physikalische Zeitschrift*, 10:75, 1908.

16. By contrast, in pre-relativistic physics $\Delta t^2$ and $\Delta x^2 + \Delta y^2 + \Delta z^2$ would have been separately invariant with respect to rotations and translations.

17. H. Minkowski. Space and time. *Physikalische Zeitschrift*, 10:75, 1908.

18. G. Ricci and T. Levi-Civita. Méthodes de calcul différentiel absolu et leurs applications. *Mathematische Annalen*, 54:125–201, 1900.

19. R. Hermann. *Ricci and Levi-Civita's tensor analysis paper: translation, comments, and additional material.* Math Sci Press, Brookline, MA, 1975.

20. These are the distances that would be measured by local observers with their rulers, or in the case of relativistic space-time, with their clocks and rulers. In the *Entwurf*, these distances are called "naturally measured."

21. In denoting the transformation, we omit indices to indicate that *all* the $y^i$ (for $i = 1, \ldots, N$) are functions of *all* the $x^i$.

22. In writing these formulas we work at the "same" manifold point, identified by the coordinates $x_0$ or transformed coordinates $y_0 = y(x_0)$.

23. E. B. Christoffel. *Über die Transformation der homogenen Differentialausdrücke zweiten Grades*, *Journal für die reine und angewandte Mathematik*, 70:46–70, 1869.

24. The Christoffel symbol is neither covariant or contravariant; rather, it "completes" a partial derivative with the exact term needed to make it a covariant object. The symbol has an intuitive geometrical interpretation for embedded surfaces, it projects the partial derivative, which may include a normal component, back into the surface. However RLC and their contemporaries had an algebraic rather than geometric understanding of covariant differentiation. Levi-Civita himself would evetually make the connection to geometry in 1917.

25. Incidentally, the metric and inverse metric can be used to lower and raise indices: $A_i = g_{ij}A^j$ is covariant, while $f^i = g^{ij}f_j$ is contravariant.

26. G. Ricci and T. Levi-Civita. Méthodes de calcul différentiel absolu et leurs applications. *Mathematische Annalen*, 54:125–201, 1900.

27. Another result with a beautiful geometrical interpretation that was not understood by RLC in 1900. See, e.g., A. Zee. *Einstein gravity in a nutshell.* Princeton University Press, Princeton, NJ, 2013.

28. More correctly, *pseudo-Riemannian* manifold, since the metric of space-time allows for negative proper distances along "time-like" directions. This is a distinction ripe with physical consequences, but one that can largely be ignored for the purpose of mathematical development.

29. Local Lorentz coordinates make direct physical sense with respect to the measurements of rods and clocks. Beyond local neighborhoods, however, coordinates lose physical meaning. Einstein initially fights, then comes to terms with this proposition—a fascinating story that is beyond our scope here. See, e.g., J. Norton, Einstein's triumph over the space-time coordinate system. *Diálogos*, 79:253, 2002.

30. I.e., point masses so light that they do not themselves affect the gravitational field.

31. A. Einstein and M. Grossmann. *Outline of a generalized theory of relativity and of a theory of gravitation.* Teubner, Leipzig, 1913.

32. The ten unique components of the metric (a symmetric 4-by-4 tensor) provide a redundant description of the gravitational degrees of freedom, and this redundancy must be removed before one can make physical predictions. One way to remove the redundancy is to impose additional conditions on one's choice of coordinates.

33. J. D. Norton. What was Einstein's "fateful prejudice"? In J. Renn, editor, *The Genesis of General Relativity*, page 715. Springer Netherlands, Dordrecht, 2007.

34. A. Einstein and M. Grossmann. *Outline of a generalized theory of relativity and of a theory of gravitation.* Teubner, Leipzig, 1913.

35. M. Janssen and J. Renn. Untying the knot: how Einstein found his way back to field equations discarded in the Zurich notebook. In J. Renn, editor, *The Genesis of General Relativity*, page 839. Springer Netherlands, Dordrecht, 2007.

36. G. Ricci and T. Levi-Civita. Méthodes de calcul différentiel absolu et leurs applications. *Mathematische Annalen*, 54:125–201, 1900.

37. See, for instance, C. W. Misner, K. S. Thorne, and J. A. Wheeler. *Gravitation*. W. H. Freeman, San Francisco, 1973, where this *modus operandi* surfaces continually.

38. A. Einstein. *Autobiographical notes*. Open Court, La Salle, IL, 1949.

39. M. Janssen, J. Renn, T. Sauer, J. D. Norton, and J. Stachel. A commentary on the notes on gravity in the Zurich notebook. In J. Renn, editor, *The Genesis of General Relativity*, page 489. Springer Netherlands, Dordrecht, 2007.

40. In the *Entwurf*, Grossmann goes as far as stating that "I have purposely not employed geometrical aids because, in my opinion, they contribute very little to an intuitive understanding of the conceptions of vector analysis."

41. G. Hessenberg. Vektorielle Begründung der Differentialgeometrie. *Mathematische Annalen*, 78:187, 1917.

42. T. Levi-Civita. Nozione di parallelismo in una varietà qualunque. *Rend. del Circ. Mat. di Palermo*, 17:173, 1917.

43. H. Weyl. *Raum. Zeit. Materie. Vorlesungen über allgemeine Relativitätstheorie*. Springer, Berlin, 1918.

44. See, e.g., A. Zee. *Einstein gravity in a nutshell*. Princeton University Press, Princeton, NJ, 2013.

45. K. Reich. Differential geometry. In I. Grattan-Guinness, editor, *Companion Encyclopedia of the History and Philosophy of the Mathematical Sciences*, volume 2, page 331. Johns Hopkins, Baltimore and London, 2003.

46. A point made, e.g., by F. Toscano, *Il genio e il gentiluomo: Einstein e il matematico italiano che salvò la teoria della relatività generale*, Alpha Test, Milano, 2004.

## Notes for Appendix B

1. It was not published. The manuscript, found among Ricci's papers, was kindly placed at my disposal by the family. [Unless otherwise indicated, all footnotes are from Levi-Civita's original memoir. The manuscript to which he alludes is now held by the Liceo scientifico statale "G. Ricci Curbastro," Lugo (RA).]

2. One need only recall the book by J. A. Schouten entitled *Der Ricci-Kalkul* (Berlin: Springer, 1924).

3. *Translators' Note*: Now called tensors.

4. E. Beltrami, "Relazione sul concorso al Premio Reale per la Matematica per l'anno 1887," *Rendiconti dell' Accademia Lincei* 5 (1889), 300–308, on pp. 304–307.

5. L. Bianchi, "Relazione sul concorso al Premio Reale, del 1901, per la Matematica," *Atti dell'Accademia dei Lincei. Rendiconti dell'adunanza solenne del 1 giugno 1902,* 2, 142–151, on pp. 147–150.

6. G. Ricci Curbastro, *Sulle condizioni idrauliche della campagna a destra del Reno-primario e sui provvedimenti atti a migliorarle,* Faenza, Conti, 1881, pp. 1–36.

7. G. Ricci Curbastro, *Nuova Relazione al Consiglio Comunale sulla Proposta di condurre a Lugo le acque delle Vallette,* Lugo, Cremonini, 1903, pp. 1–41.

8. *Translators' Note*: There is no record of this usage in English; in Italian it meant the parts of algebra that were most useful for analysis.

9. G. Ricci-Curbastro, *Lezioni di analisi infinitesimal: funzioni di una variabile,* Padua, 1926.

10. "Sul concetto di successione in relazione col teorema fondamentale del calcolo integrale," Atti. R. Ist. Veneto, 69(1910), p. 1055.

    *Translators' Note*: It is not altogether clear why Levi-Civita singles out this exposition (published in Ch. 2 of the 1926 textbook) as especially original, given that Ricci follows quite closely the standard definition of the Darboux integral. The derivation does however embody the principle, dear to Ricci, of the "unity of the calculus around the fundamental concept of sequence" (Ricci 1926, in the unfinished preface); this may be the reason why Levi-Civita would consider it especially characteristic of Ricci.

11. *Translators' Note*: We believe this is referring to the famous Ricci-TLC paper published in 1900.

12. *Translators' Note*: The two contracted Riemann tensors referred to here are the contraction $R_{ij}$ of the Riemann tensor $R_{ij}{}^{k}{}_{l}$ for any Riemannian manifold by Ricci (1904), while the second one is the contraction of the Riemann tensor for any 4 dimensional *semi-Riemannian* manifold of signature (1,3) (i.e. space-time) by Einstein and Grossman (1913). Since Ricci's tensor was defined nine years earlier than Einstein and Grossman's, this tensor is now almost always referred to as the Ricci tensor (or curvature) for any semi-Riemannian manifold.

13. *Translators' Note*: The *Einstein tensor* is defined by $G_{ij} := R_{ij} - kg_{ij}R$ where $R$ is the *scalar* Ricci tensor, $g_{ij}$ is the Lorentzian pseudo-metric on space time, and $k$ is a suitable constant.

14. *Translators' Note*: In the usual inertial coordinates, $ds^2 = dx^2 + dy^2 + dz^2 - dt^2$.

15. Friedrich Kottler, "Über die Raumzeitlinien der Minkowski'schen Welt," Kaiserliche Akademie der Wissenschaften (Vienna), Mathematisch-naturwissenschaftliche Klasse, Abteilung IIa. Sitzungsberichte 121 (1912): 1659–1759.

16. *Translators' Note*: Usually written as $ds^2 = \sum_{i,j=1,2,3,4} g_{ij}dx^i dx^j$, where the $\{g_{ij}\}$ are the unknowns.

17. Albert Einstein and Marcel Grossmann, *Entwurf einer verallgemeinerten Relativitätstheorie und einer Theorie der Gravitation,* Leipzig, 1913.

# Index